THE SCIENCE OF MICHAEL CRICHTON

Other Titles in the Science of Pop Culture Series

The Science of Dune (January 2008)

BENBELLA BOOKS, INC.
Dallas, Texas

BenBella Books, Inc.
6440 N. Central Expressway, Suite 503
Dallas, TX 75206
www.benbellabooks.com
Send feedback to feedback@benbellabooks.com

Printed in the United States of America
10 9 8 7 6 5 4 3 2 1

Library of Congress Cataloging-in-Publication Data

The science of Michael Crichton : an unauthorized exploration into the real science behind the fictional worlds of Michael Crichton / edited by Kevin R. Grazier.
p. cm.
ISBN 1-933771-32-1
1. Crichton, Michael, 1942—Criticism and interpretation. 2. Science fiction, American--History and criticism. 3. Science in literature. I. Grazier, Kevin Robert, 1961–

PS3553.R48Z78 2008
813'.54—dc22

2007041419

Proofreading by Emily Chauviere and Stacia Seaman
Cover design by Laura Watkins
Cover illustration by Ralph Voltz
Text design and composition by John Reinhardt Book Design
Printed by Bang Printing

Distributed by Independent Publishers Group
To order call (800) 888-4741
www.ipgbook.com

For special sales contact Robyn White at robyn@benbellabooks.com

CONTENTS

INTRODUCTION

PERHAPS IT IS UNBECOMING TO GUSH, but I have to admit up front that I am a huge fan of Michael Crichton's writing and have been for most of my life. I've never met the man but, then again, they say you should never meet your heroes. Like many of his readers I have been thrilled, terrified, stimulated, and certainly entertained by his novels. When a new Michael Crichton novel is released, I'm the first in line. Although movie adaptations rarely do his novels justice, *Andromeda Strain* and *Jurassic Park* still rank among my all-time favorites. Further, I would highly recommend any/all of the candid essays and speech transcripts on his Web site (www.michaelcrichton.com) to anybody interested in the state of science today. Many of his postings are true eye-openers, particularly for the non-scientist. On the other hand, having, in fact, read all the content posted on his Web site, I know that he and I would vehemently disagree on several topics—whether one of the courses I teach at UCLA, the Search for Extraterrestrial Life, is even worthy of being classified as a science class, for example. I expect he would argue that my class belongs in the "Pseudo-Scientific Philosophy Courses" section of the university catalog. Okay then, so maybe "hero" was a slight overstatement. Still, not only have I thoroughly enjoyed Crichton's works, they've spoken to me in very different ways over the span of both my life and career—often in pertinent and timely fashions.

In the mid-1970s, when I was a high school student and scientist-wannabe growing up in the blue-collar suburbs of Detroit, I discovered the movie—and later the book—*The Andromeda Strain*. That story gave me my first multiple epiphany! For the first time in fiction I saw scientists as real, flawed, people, not simple caricatures. Moreover, although I had been a science fiction fan from very early on, and as much as I credit the original *Star Trek* for its role in starting me on the road to a career in science, the whole *Trek* universe and its "can't we all just get along" view of the future, had a... sanitized... feel to it. *Andromeda Strain* said very clearly that, as Humans venture into space, we might encounter monsters more horrible than those with sharp teeth, acid blood, or laser blasters, and they may very well be microscopic. Stephen King once said, "I recognize terror as the finest emotion and so I will try to terrorize the reader. But if I find that I cannot terrify, I will try to horrify, and if I find that I cannot horrify, I'll go for the gross-out. I'm not proud." Anybody who has seen a horror movie within the past three decades has witnessed all manners of gruesome and bloody dispatch. Are any of Jason's murders truly any worse than having all the blood in your body clot over the span of a few seconds? Imagine having a massive heart attack, a massive stroke, and terminal atherosclerosis simultaneously. So, then, from *Andromeda Strain* I also learned—something the early horror author H. P. Lovecraft relied upon long ago, and what many horror writers of today seem to have forgotten—that a truly terrifying work of fiction engages in a way that makes the reader's mind do the work, more than the mere anticipation of the "gross-out."

If you took a poll among scientists, irrespective of discipline, to find what initially attracted them to science, I would bet good money that two responses would overwhelmingly outnumber all others: space and dinosaurs. I managed to squeeze in a reading of *Jurassic Park* during Christmas break while still a graduate student at Purdue University. Coincidentally, I had just TAed an undergraduate course on dinosaurs the previous semester. With undergraduate degrees in computer science and physics in hand, I was looking forward to a career in the Earth and planetary sciences and excited about the prospect of doing interdisciplinary research. It seemed to me at the time that real progress in science is made when the techniques of multiple disciplines are brought to bear on a problem. At the same time, I was not looking forward to several years as a poor graduate student. Enter *Jurassic Park*—a multidisciplinary mixture of supercomputing, genetic sequencing, chaos, and, yes, dinosaurs. A graduate student in the sciences must derive motivation from multiple sources, big and small, to endure the long haul. It is no exaggeration that, for me, the inherent coolness of the novel *Jurassic Park*—as well as the fact that it addressed so many topics I found interesting—was one of many such nudges. In fact, there is a well-written two-and-a-half-page chapter in *Jurassic Park* entitled "Destroying the Planet" which I still read aloud to many of my college classes to this day.

As a mid-career scientist, I still find Michael Crichton's writings as pertinent to my life as ever, and find that they often echo my own sentiments—which likely explains why I'm the first person in line at the bookstore when he writes a new novel. Putting aside, for the moment, Crichton's specific critiques—technological and sociological—of global warming (which are amply covered in this volume), *State of Fear* provides excellent commentary on several aspects of how unscientific the scientific process can be today.

When a graduate student picks a dissertation advisor, he or she has often unknowingly made a *de facto* choice of sides in a scientific debate—whatever one(s) in which the advisor is embroiled. When I chose my dissertation advisor at UCLA, I was immediately enmeshed in a "holy war" over computational techniques now used within the celestial mechanics community—a conflict that still rages, but one which Moore's Law will ultimately decide, in all likelihood. I recall one particular grant proposal rejection letter that my advisor received. He had proposed doing research using a particular methodology, but not the one "in vogue" with the bulk of the community, and the rejection was filled with more not-too-thinly-veiled *ad hominem* attacks than scientific arguments and/or justifications. It said, in short, "You just aren't part of the 'in crowd.'" It was eye-opening disillusioning, a dose of the "real world," and apparently not unusual, given Crichton's appendix in *State of Fear*. In it, Crichton draws many sobering parallels between the environment that existed around the eugenics movement of the early twentieth century, and that of global warming and climate change research of the early twenty-first: there is no debate here, you are to be marginalized if you aren't part of the "in crowd." As science advisor on *Battlestar Galactica*, I should point out that we have a saying for this, "All of this has happened before, and all of this has happened again." Other examples of similar unscientific behavior from scientists exist: do an Internet search on the Palmdale Bulge when you have an hour or so to kill.

Crichton's novel *Next* is another cautionary tale on the state of the biotech industry today. In recent books like *State of Fear* and *Next*, his online essays, and his speeches, Michael Crichton does an excellent job of pointing out where science today can be far more . . . scientific. As much as I dislike the trendy saying, Crichton is all about "keeping science real." Some scientists appreciate it, some even pay attention, and again I find his writings speak to me in pertinent and timely ways.

It will be no surprise, then, when I say that I was thrilled when asked to edit an entire book of essays dedicated to exploring the scientific aspects of Michael Crichton's novels. It never ceases to amaze me how much research is put into his novels and, although ostensibly science fiction, how much real science the novels contain. The following essays examining his work are all very well-written in my opinion. I'll admit up front that I do not agree 100 percent with every single scientific viewpoint shared in the essays, but then again I don't have to. That's what the process of scientific debate is about. I hope Michael Crichton would approve.

Oh, and Dr. Crichton, if you're reading this...I'm waiting for that next book.

—KEVIN R. GRAZIER, PH.D.
December 2007

THE ANDROMEDA STRAIN

Sergio Pistoi

On 28 September 1969 the Murchison meteorite fell in Victoria, Australia. Within that meteorite were compounds related to sugars, and more than seventy amino acids—the building blocks of earthly life—fifty of which are not present on Earth. On the Space Shuttle Atlantis *mission STS-115, twenty-seven years later,* Salmonella typhimurium *bacteria showed a dramatic increase in virulence as a result of space flight. Sergio Pistoi, Ph.D., examines* The Andromeda Strain *which, suddenly, seems really quite plausible.*

This book recounts the five-day history of a major American scientific crisis.

As in most crises, the events surrounding the Andromeda Strain were a compound of foresight and foolishness, innocence and ignorance. Nearly everyone involved had moments of great brilliance, and moments of unaccountable stupidity.

—Michael Crichton, foreword to *The Andromeda Strain*

LATE 1960s. A secret military satellite, sent into deep space by the army to look for new forms of life, crashes near a small village in Arizona, spreading a mysterious and deadly extraterrestrial organism. People coming into contact with the organism die instantly, because the bug causes all their blood to clot solid. It's something nobody has ever seen on Earth. Something that could potentially destroy humanity if it spread.

The Andromeda Strain recounts the drama of four scientists fighting against that lethal creature from space. Researchers start by exploring the village and by collecting samples of the mysterious organism, while the army isolates the area. They still have no idea of what they are facing, and they don't know how to stop that deadly invasion. But they have found two survivors—an old, alcohol-addicted man and an always-crying newborn baby. Whatever the old man and the baby have in common, it must hold the key to understanding how the deadly organism works and how it can be stopped.

Michael Crichton published *The Andromeda Strain* in 1969, when he was still a graduate student at the Harvard Medical School. It was his first bestseller, making him known worldwide as an emerging literary star. The story has all the ingredients of a great bio-thriller, and at the same time reflects Crichton's medical and scientific background and his ability to mix real science with creative, but plausible fiction. Crichton wrote the novel in a false-document style, mixing references to real scientific publications and to fictional "classified" documents, like he was recounting real events. Also, the portraits of researchers are quite realistic in the way they collect data about the organism (codenamed *Andromeda*, hence the title) and use scientific knowledge and intuition to formulate hypotheses about its nature. It's also interesting that, apart from the inevitable science fiction touch, much of the equipment and methods described in the book reflects the best technology available during the 1960s.

Consistently with its extraterrestrial nature, Andromeda is astonishingly different from any known form of life; it has no proteins, no DNA, and none of the other building blocks that are typical of terrestrial organisms. Surprisingly, it has the structure of a crystal, something that we associate with inorganic objects such as minerals, not with a living organism. If judged with terrestrial criteria, Andromeda cannot be living. Still, it behaves like a living organism: it is able to reproduce and to infect its victims and, as it will turn out later in the story, it can mutate very rapidly, adapting to its environment.

How plausible is this scenario? Based on scientific evidence, can we imagine alien organisms like Andromeda, so far from our idea of life? If so, how could we protect the Earth from the invasion of alien pests? As weird as they sound, these questions are not just science fiction buffs' speculations, but also the subject of many serious and fascinating scientific investigations. But first of all, before you laugh at the idea of a living crystal, ask yourself a question: What do you *exactly* mean by "life"?

A Matter of Taste

Surprisingly, a simple question like "What is a living organism?" has no clear-cut answer. All attempts to scientifically define what is and isn't alive have failed miserably. Of course, we can list some features that are closely associated with life: living organisms are able to reproduce, passing their genes to their offspring; they constantly maintain their internal equilibrium, even when their environment changes (a property that biologists call homeostasis); and they have a metabolism that allows them to transform and store energy. However, none of these features alone can define life beyond reasonable doubt. For example, not all the individuals of a species reproduce. I don't have children, but I was alive last time I checked. On the other hand, many objects are capable of homeostasis and are able to transform energy. Imagine what would happen if an alien landed on a farm while looking for signs of life on Earth and found only a refrigerator and a bag of seeds. Which of the two would look more "alive"? A refrigerator is capable of homeostasis (because its thermostat actively maintains a constant internal temperature and humidity) and it transforms energy, while the seeds are as inert as stones. You could argue that a refrigerator does not reproduce, but neither do the seeds in the bag, unless they are planted and kept in the right conditions to germinate. In terms of pure evidence, and according to our criteria, the alien would understandably dismiss the seeds as objects, asking to be taken to the refrigerator's leader instead.

But we all know that the seeds, not the refrigerator, are the living stuff in the room. So where's the catch? There is no catch. We can tell that the seeds are living organisms because *we know* they are part of a biological cycle: although they look inanimate, we know that they have a potential to germinate, giving rise to a plant. Likewise, we say that a refrigerator is not an organism because *we know* it is a human artifact without any life on its own and without a potential to reproduce. In other words, we cannot define life a priori, but we can recognize it based on what we know about life on Earth.

Our definition of life is so blurred that, eighty years after their discovery, scientists are still debating whether viruses can be considered alive or not. Strictly speaking, viruses are unable to meet the minimum requirements by which we define life: they are sorts of genomic capsules made of proteins and containing a payload of DNA or RNA. Since they cannot read and copy their own genome, like malicious software viruses they must infect other cells and harness their machineries to produce new viral copies that, in turn, infect other cells. Because of that, viruses are molecular parasites without life on their own, living a "borrowed life" inside other organisms. Are they alive or not? Back in 1962, the French Nobel prize laureate André Lwoff put the question in very frank terms: "Whether [viruses] should be regarded as living organisms is a matter of taste," he wrote. His famous quote is still very valuable today.

If simple, terrestrial organisms like viruses already are a challenge to our concept of life, how could we recognize alien organisms based upon a different biology than ours if they existed in another planet? It's not a simple question. One problem is that we are so accustomed to our biology that we can hardly conceive a different one. Terrestrial life is like a set of Lego toys: with only a small set of building blocks and a few, simple rules, evolution has given rise to an infinite variety of shapes, species, and functions. From a coliform bacterium to Angelina Jolie, the basic chemical recipe is the same: four elements—carbon, hydrogen, oxygen, and nitrogen (the so-called CHON group)—make up 99 percent of the mass of every living creature, while thousands of other elements, including calcium, phosphor, and sulfur, account for the remaining 1 percent. Carbon provides the backbone of all organic molecules and is therefore the basis for the chemistry of life on our planet. In cells, carbon atoms are combined with the other elements to form organic molecules such as amino acids, sugars, fats, and nucleotides, which, at their turn, are the building blocks of bigger molecules such as proteins, DNA, or cellular membranes. Water, which is abundant on Earth, provides the solvent and the chemical environment for all biological reactions. The beauty of the system is that, just like a set of Lego toys, the building blocks are essentially the same for all creatures: by combining, in different ways, twenty amino acids, five nucleotides, and a handful of other molecules, you can obtain every organism on Earth, whether it is a bacterium, a banana, or your neighbor. This scheme works well for our planet; however, astrobiologists warn that it's just one of the many possibilities by which life could have evolved in the universe. On another planet, biology could follow a completely different paradigm, for example, by using a different set of building blocks and different rules for their assembly.

A Crystalline Lifestyle

Even using very loose criteria to define life, it's hard to imagine how a crystal, like the Andromeda strain, could behave as a living organism. A crystal is almost the nemesis of our idea of life: a fixed, dry structure, with few or no possibilities of dynamic changes. Take, for example, the crystalline form of carbon (in the "chair" hexagonal structure, better known as a diamond). Girls' best friend is a regular network of carbon atoms kept together by sturdy chemical links: they are beautiful and incredibly solid, but, chemically speaking, they are a death valley with no or few possibilities for reactions. Under these conditions, how could a crystal *live*?

Surprisingly, although there is no evidence of them on this planet or elsewhere in space, the existence of crystalline organisms is not just a science fiction's expedient, but a possibility that some researchers have seriously considered. One of them is Scottish biologist Alexander Cairns-Smith, the author of the popular and controversial book *Seven Clues to the Origins of Life,* who suggested in the 1960s that crystals, and not organic molecules, were the earliest forms of life on Earth. His theory was based on some interesting features of crystals: they are able to grow (you probably enjoyed the beauty of growing crystals during your school science classes), and, what is intriguing, when two pieces of crystal come apart and grow further, they maintain their shape, mimicking a sort of reproduction. Besides, many minerals are good catalysts for biological reactions. Cairns-Smith postulated that crystals would have been good candidates as primitive forms of "life" that could have later evolved into more complex organisms because of these and other properties. According to his "genetic takeover" model, inorganic crystals (probably clays) provided the first self-replicating and evolving entities on Earth and only later in evolution did RNA and DNA take over as a more efficient hereditary material for living organisms. Like other theories about the origin of life, Cairns-Smith's model is difficult to prove and has its share of enthusiasts and critics. Although bizarre, the idea is neither more nor less plausible than others: the origin of life is another field where choosing among different explanations is still largely a matter of taste. The Scottish scientist, however, is not alone in believing that a non-chemical life could be possible. In 1976, French astronomer Jean Schneider wrote a scientific paper proposing that non-chemical, crystalline life would be possible, at least in theory. His model of "crystalline physiology," as he called it, was based on the existence of tiny defects, or interruptions, in the regular atomic structure of the crystal, called dislocations. You can think of them as cracks on a glass windshield. The analogy may not be very accurate, but it is useful to visualize Schneider's theory. Vibrations make the cracks slowly propagate along the windshield and when two of them intersect, they produce new patterns. Likewise, dislocations in a crystal propagate and interact, producing new dislocations with mathematically predictable patterns, mimicking a simple reaction. Like a crack on a glass, every dislocation has a unique shape, which you can see as a form of information, like a drawing. When a crystal grows and comes apart, dislocations are passed to the "daughter" crystals, a bit like genetic information. Schneider speculated therefore that dislocations could work as a primitive "central memory"; the equivalent of DNA in a hypothetical crystal world. Since solid crystals don't move, the French astronomer also suggested that liquid crystals, which flow like a liquid, but have atoms arranged in a crystal-like way, would be a more likely candidate for life.

Whether or not you believe in the hypothesis of a crystal ancestor, Cairns-Smith's theory offers a great opportunity to speculate about the origin of the Andromeda strain. The oldest fossils found so far are about 3.5 billion years old; therefore, if Cairns-Smith were right, life as we know it today should have evolved from crystalline "organisms" sometime between 3.9 and 4.1 billion years ago. Since we haven't found any crystalline organism on Earth, we can assume that carbon-based organisms outperformed crystal species so well that all living crystals, if they ever existed, are now extinguished. However, moving back into the world of Michael Crichton, we can think of another fascinating possibility: imagine for a moment that crystalline life did not extinguish, but instead, in a remote and inaccessible part of the world, evolution continued to produce more and more sophisticated crystalline species and Andromeda was one of them. Instead of a space bug, Andromeda could be a far relative of ours, a descendant of the early crystalline forms of life on Earth, a living fossil of a forgotten branch of evolution that stuck with crystalline, rather than carbon, organization. I like to think that Crichton would have mentioned this possibility in his book, had he been aware of Cairns-Smith's work. He probably missed that chance by a hairsbreadth: Cairns-Smith wrote his first paper on the subject in 1966, only three years before *The Andromeda Strain* was published. Unfortunately, it was just a minor article in an obscure scientific journal that went overlooked for many years.

Look, No Enzymes!

If we have to imagine a hypothetical world populated by crystal creatures, we must explain how they would be able to use and transform energy to grow, to build their components, and to maintain their homeostatic balance. In our biology, these activities depend on a series of chemical processes, which are known collectively by the name of *metabolism*. Catabolic and anabolic reactions are the Yin and Yang of metabolism: the former produce energy by breaking down nutrients, while the latter use energy to build all the cell's components. You can think of the cell as a factory where chemical reactions are organized into assembly and disassembly lines that biologists call *pathways*. Catabolic pathways are organized as a series of disassembly lines in which energy-rich carbon compounds, typically sugars and fats, are broken down to produce energy. The most familiar catabolic activity is respiration, in which cells break down nutrients into water and carbon dioxide, which, in vertebrates, is eventually expelled from the lungs or the gills.

Anabolic pathways, instead, are assembly lines where all the various building blocks are synthesized and rearranged to produce DNA, proteins, membranes, and all the components of a cell, and to build up storage, mainly in the form of long chains of sugars or fat.

If the cell is a chemical factory, enzymes are the workers that make everything happen. Our genome codes for thousands of enzymes, each specializing in one step of a pathway, like workers on a chain line. Enzymes are proteins that catalyze (i.e., accelerate) chemical reactions.

Without them, the whole cell would be blocked like a factory in the middle of a strike. Metabolic diseases are a dramatic example of the importance of enzymes. In these disorders, a single, faulty enzyme can block an entire pathway, depriving the organism of essential components, while unfinished building blocks accumulating engulfing the cells and leading to severe symptoms or death.

Life as we know it would be impossible without enzymes. However, this does not mean that enzymes are the only possible biological catalysts. In fact, we can imagine one or more alternative worlds in which different compounds would substitute protein enzymes. Metal ions such as manganese, nickel, zinc, cobalt, and many others are good catalysts for many biochemical reactions. In fact, they are so good that many enzymes use them as co-factors: while the protein part, called apoenzyme, dictates the rate and specificity of the reaction, the metals perform the catalysis. An organism deprived of proteins, like Andromeda, could rely on such metals as catalysts, provided that they are coupled with more complex compounds acting as apoenzymes. For example, metal could couple with hydrocarbons, which are abundant in the universe and, like proteins, can form complex structures.

Aliens in My Steak

Like all people from Tuscany, I have an idiosyncratic approach to the perfect *bistecca alla fiorentina.* The meat must come only from Chianina beef, the huge oxen bred in Val di Chiana, near Arezzo; it should be at least 5 cm thick and, of course, with its T-bone. The authentic *fiorentina* should be deliciously crusty outside but raw and juicy inside: the trick is to cook it only on a wood grill, keeping it very close to the coal, for no more than four minutes on each side, and turn it only once. Also, if you want to keep it tender, add salt only after cooking.

The *fiorentina* is such an institution for Tuscans that at the beginning of 2001, when the mad cow scare was ravaging the beef industry worldwide, Chianina beef was still selling well in the region.

My countrymen were undeterred by mad cow, but, to their horror, the EU authorities decided to ban the *fiorentina* anyway, together with all T-bone steaks, fearing that the nerve tissue in the bone (which is actually part of a cow's vertebrae) could potentially transmit the disease to humans. The night before the ban took effect, many Tuscan cities organized "farewell" feasts where people enjoyed their last juicy slices for a while.

The outbreak of *bovine spongiform encephalopathy*, or BSE (the scientific name for mad cow disease), exploded in Britain's cattle in the mid-1980s, almost certainly because of the practice, now forbidden, of feeding animals with already infected meat and bones. The alarm bell for human health rang in the 1990s, when scientists discovered that a rare form of human spongiform encephalopathy, known as variant Creutzfeldt-Jakob disease (vCJD), was due to the same infectious agent that caused BSE, most likely transmitted by eating the meat of affected animals.

The term "spongiform" is appropriate to describe a horrifying effect of the disease: the brains of affected animals and patients become like a sponge, with holes corresponding to parts where neurons have disappeared, leading to a progressive loss of coordination, neurodegeneration, and, eventually, death.

The fear of a devastating epidemic, however, was only a part of the BSE story. While panic was spreading globally, wreaking havoc on the meat industry and sentencing my beloved *fiorentina* to quarantine, something else about that disease was sending shivers down the spine of biologists. BSE was not caused by the usual suspects, viruses or bacteria, but, instead, by something new and weird—an infectious pathogen that was so far from our idea of life that it almost looked like a space creature. That thing gave a blow to our received beliefs on biology, and its discovery was one of the most exciting and disturbing biological revelations of the century.

In 1982, American neurologist and would-be Nobel laureate Stanley Prusiner confirmed what many scientists had already suspected: the infectious agent causing BSE was neither a bacterium nor a virus: it was a protein, without any DNA or RNA. The idea of an infectious protein was so unfamiliar that Prusiner had to coin a new name for it: he called it *proteinaceous infectious particle*, or, abbreviated, *prion*. Many scientists were skeptical. How could a protein replicate without any genetic material? How could it propagate and infect other organisms? This idea seemed as strange as that of a space bug. It took Prusiner more than a decade before he could convince his colleagues about the existence of prions and demonstrate their mechanism of action. Surprisingly, it turned out that the infectious prions were mutant forms of innocuous proteins called PrP (for PRion Protein) that exist normally in most species, from yeast to humans. Prusiner also discovered that the mutated forms of prions had a slightly different three-dimensional shape than their normal counterpart. When the mutant and normal prions came into contact, the former altered the shape of the latter, transforming it into a mutant. The whole process reminds one of a zombie B-movie: bad prions transform normal ones into mutants, which in turn attack other normal PrPs, resulting in a frightening chain reaction. Prion diseases, like BSE, happen when mutant prions enter the brain and come in contact with the normal PrP in the neurons, triggering this slow but inexorable zombie mechanism. Because of their altered shape, mutant prions tend to stick together, forming bulky lumps that engulf neurons and kill them.

Besides vCJD, transmissible human prion disorders include kuru, a disease discovered in the 1950s in an isolated tribe in Papua New Guinea and now almost disappeared. Fatal familial insomnia (FFI) and Gerstmann-Straussler-Scheinker disease (GSS) are two hereditary forms that affect only a few families around the world and are not because of infection by mutant prions, but instead, affected people produce abnormal prions as a result of mutations in their DNA.

Although the exact role of normal prions is still unknown, their composition is similar to a class of protein called *chaperones*, whose function is to protect other proteins and preserve their shape. It's ironic that, despite their function, prions cannot protect themselves from the attack of their mutant counterparts.

The existence of prions is another challenge to our received ideas about life, bigger than that caused by viruses eighty years ago. Both viruses and prions are molecular parasites that need to infect other cells to replicate. However, unlike viruses, prions lack a genome, the lowest common denominator of all known organisms on Earth. Whether you want to see them as a form of life or not is again a matter of taste. "*Life* and *living* are words that the scientist has borrowed from the plain man," British scientist Norman Pirie wrote in the 1930s, referring to viruses. "[N]ow, however, systems are being discovered and studied which are neither obviously living nor obviously dead, and it is necessary to define these words or else give up using them and coin others." Today, after the discovery of prions, his words sound more prophetic than ever.

The Worst-Scenario Hypothesis

When planning the first manned mission to the moon, NASA considered the risk of an alien invasion on Earth quite seriously. Although the biggest concern for scientists was to make sure that the lunar soil collected by the astronauts was not contaminated by terrestrial organisms, they also took measures to avoid the eventuality that unknown extraterrestrial microorganisms could be spread into our ecosystem. After splashdown, and before exiting the module, Apollo 11 astronauts wore special plastic coveralls and the lunar rock samples were closed into special isolating containers. Both the astronauts and the samples were then flown right away to the Lunar Receiving Laboratory (LRL), a quarantine facility that NASA built at the Johnson Space Center in Houston, Texas. The LRL included an Astronaut Reception Area where the crew was isolated and medically monitored for twenty-one days, and a Lunar Sample Laboratory, equipped with special glove-box vacuum chambers, where the moon specimens were kept under isolation and tested on dozens of plants and animal species to make sure that they did not contain dangerous pathogens.

Unfortunately, if Apollo had brought back any microorganism from the moon, all these measures would have failed to protect the Earth from contamination. A 2003 report from the Committee on Planetary and Lunar Exploration (COMPLEX) of the National Research Council concluded that, despite NASA plans, at least two places would have been likely contaminated: the Pacific Ocean, at the site of splashdown, due to poor isolation of the module, and the LRL facility itself. The report mentions a chilling episode: one day during December 1969, while examining Apollo 12 lunar rocks, technicians discovered a cut in one of the LRL glove chambers, which could have exposed workers to contact with contaminated air. Exposed personnel were sent to quarantine, but, according to the report, some escaped the facility before they could be forced into isolation. We can only imagine what would have happened if these people were contaminated by a dangerous and unknown extraterrestrial pathogen. According to the same report, NASA's worst-case scenario guidelines stated "the preservation of human life should take precedence over the maintenance of quarantine." Thus, if a fire broke out in the LRL, or if an astronaut should be evacuated from the facility because of a medical emergency, the plan, at least officially, was to break quarantine, with the risk of producing a contamination. Evidence of life on the Moon was never found, and NASA lifted the quarantine for astronauts by the Apollo 15 mission. All's well that ends well, although NASA's approach to extraterrestrial risk at the time was probably not better that the "compound of foresight and foolishness, innocence and ignorance" that, in the words of Crichton, characterized the events surrounding the Andromeda strain.

When conceiving Wildfire, the fictional quarantine plant where Andromeda was being analyzed, Crichton was clearly inspired by the NASA's Lunar Receiving Laboratory, and it is probably not a coincidence that *The Andromeda Strain* was released on May 12, 1969, only a few months before the launch of the Apollo 11 mission. Of course, Wildfire had an inevitable science fiction touch compared to its real counterpart: while the LRL looked like any other building (it was, and still is, hosted in a section of building 37 at the Johnson Space Center), Crichton imagined Wildfire as a secret underground plant with five floors, each isolated from the others and each equipped to deal with progressively risky organisms. To reach the bottom level, which had the highest security level, scientists had to undergo a series of decontaminating steps, including radiation treatments and drugs that eliminate microorganisms in the skin and the intestine. Incidentally, many of these treatments would be possible in reality only at the cost of killing people, since our body relies on bacteria to protect the skin and the mucosa from dangerous germs, help with digesting food, and produce compounds, such as vitamins, which we are not able to synthesize. In the worst-case scenario, Wildfire was also programmed to self-destroy with its occupants by blasting an automatic atomic charge, whereas NASA, as we have seen, had more humane plans (at least officially).

The question of how to keep space invaders at bay reemerged when many space agencies started planning return trips to Mars. The European Space Agency (ESA) is developing a mission to bring back Martian soil samples starting in 2013, and NASA is considering a similar endeavor for around 2020. The Russian and Chinese space agencies are also working on a return mission to the Martian moon Phobos. While there is a possibility that life on Mars may have existed, or still exists—an aim of those missions is precisely to look for possible evidence of that—biologists agree that the chances of finding dangerous germs in another planet are actually quite small. One reason for that is straightforward: infecting other organisms is not an easy task for germs. To be successful, they must elude many defensive layers deployed by the victim, and if they survive, they must have a molecular key to enter the host's cells. A parasite such as a virus must also be able to use the cell's machinery to replicate itself. Like old couples, terrestrial germs and their victims share the same biology and have evolved in parallel for millions of years, learning to "know" each other and, in some way, becoming interdependent. In this mechanism, which biologists call co-evolution, natural selection favors the germs that are better at replicating and infecting their hosts, but, conversely, also the organisms that are more resistant to infection: the result is a continuous arms race in which germs have evolved more and more sophisticated strategies to infect their victims, often at the cost of becoming specialized only in one species. An alien microorganism coming out of the blue would hardly be able to attack terrestrial organisms, escape their natural defenses, and enter their cells (providing that it would be interested to do it in the first place), especially if its biology was very different than ours. A more foreseeable scenario is that, although innocuous for our health, an alien organism could accidentally find our habitat so favorable that it would grow and proliferate without control, altering the ecosystem. Something similar happened when European rabbits were introduced in Australia in the mid-1800s: once there, they found ideal conditions for reproduction and no or few natural predators. The likely descendant of only twenty-four individuals brought from the old world caused an invasion of millions that are responsible for the destruction of habitats and the extinction of many indigenous species, besides huge damage to agriculture.

To limit the possibility of alien invasions, the COMPLEX report recommends treating Martian samples as if they contained the most dangerous organisms on Earth. The receiving facility, for example, would include BSL-4 laboratories, the most stringent level of biological containment available today. In these laboratories, which are used to study lethal, airborne germs like the Ebola virus, samples are kept under sealed glove cabinets and the room itself is totally isolated from the external environment. Moreover, the air is filtrated and kept under a lower pressure than outside so that it will not exit the room even if a door is opened. Inside BSL-4 labs, operators can also wear spacesuits fitted with a respirator when they need to manipulate the samples closely. As a further protection, the report recommends building an additional isolating perimeter around the BSL-4 laboratories.

These precautions could work well once Martian samples have reached the facility. However, there is a possibility that an accident happening before reaching the destination could spread Martian microorganisms into our biosphere. An example of that occurred in 2004, when *Genesis*, a NASA space probe, returned to Earth after spending twenty-seven months in deep space collecting solar wind particles. NASA had planned for an ultra-soft landing. The capsule containing the precious space samples was supposed to be slowed down by a parachute and caught midair by a helicopter flown by a stunt pilot. Unfortunately, the parachute failed to deploy properly and the capsule shattered at high speed in the Utah desert, exposing the extraterrestrial samples to utterly terrestrial dust. The *Genesis* samples were not dangerous, and scientists could eventually salvage most of them to use in their research, but a similar accident occurring to a spacecraft returning with a load of Martian samples would be a potential nightmare that engineers are determined to avert. Although there are many imaginative solutions on the drawing board, including that of shipping Martian samples directly to a receiving laboratory on the Moon (or ISS), space agencies will probably opt for the most simple and straightforward way: putting Martian samples into crash-proof containers, strong enough to resist a bad landing such as the one that destroyed the *Genesis* probe.

If and when missions to the red planet will return to Earth, they will not be the first to bring Martian material into our environment, anyway. According to some estimates, half a ton of Martian material has fallen onto the Earth every year since the birth or our planet in the form of meteorites. The Jet Propulsion Laboratory in Pasadena, CA, lists thirty-four Martian meteorites discovered so far on Earth. Whether any of them contained living organisms is still hotly debated. In 1996, a team from NASA hit the headlines when they claimed to have discovered evidence of micro-fossils in one Martian meteorite called ALH84001, found in Antarctica; however, most experts are still skeptical and the question is still open after more than a decade. If there were any unequivocal evidence of extraterrestrial life in a Martian meteorite, this would open new and fascinating scenarios about life in the universe, and the origin of life on our planet.

An Acid Happy Ending

Let's go back to the puzzle that the Wildfire scientists are trying to solve. Only two people have survived after being infected by the Andromeda strain: one is an always-crying newborn baby, the other is an old man addicted to *sterno*, or "*canned heat*"—a mixture of ethanol and methanol used as a combustible but also as an off-label alcoholic drink. What could an old drunk and a screaming baby have in common, and why did they both survive the deadly infection?

The answer is simple, although it comes only after many failures and sidetracks: Andromeda is a hard-boiled creature adapted to thrive in the ultra-severe conditions of deep space, but once in the blood of victims, it can live only within a narrow range of pH, corresponding to that of normal blood. Honestly, it's difficult to imagine how a creature that has survived in the harshest conditions of outer space could be so sensitive to acidity: let's say it sounds as plausible as the Terminator being afraid of a snowball. Everybody has a soft spot anyway, and Andromeda's pickiness, together with some basic knowledge of blood physiology, will be the key to Crichton's happy ending. Acidity is measured by a number called pH, whose values range approximately between -5 and +14. Pure water has a pH of 7.0, which is considered neutral. Values of pH below and over 7.0 are acid and basic (or alkaline), respectively. Human blood is slightly alkaline (its normal pH is between 7.35 and 7.45) and a number of physiological mechanisms help to keep this balance carefully within limits. Hyperventilation is a simple way to alter the acidity of your blood: if you stay still and breathe faster than normal, your lungs will eliminate more carbon dioxide (CO_2), a weak acid, from the blood. As a result, the blood will become more alkaline (doctors call this phenomenon alkalosis), and after a few moments of rapid breathing, the brain reacts to this change by making you feel dizzy and lightheaded. A prolonged hyperventilation usually provokes fainting: you can see it as a way by which the brain reacts to alkalosis, restoring a normal breathing. By screaming incessantly, the baby found by Wildfire scientists hyperventilates, maintaining a constant state of alkalosis, thus his blood becomes too alkaline for Andromeda's taste. By contrast, the old man's blood is too acid, due to his habit of drinking sterno, which, among other unpleasant effects, produces acidosis. Drinking canned heat may not be the healthiest of habits, but, at least in our case, it helped to keep aliens at bay. Incidentally, the two survivors will be saved even after their blood has returned to normal, because, by that time, Andromeda mutates into a non-lethal form, which attacks objects (namely, rubber) instead of humans.

Andromeda's Achilles' heel is one thing that Crichton has probably taken from his medical background: as a doctor, he knew that the pH is a very critical issue for the survival of most germs. Microorganisms on Earth can only thrive within a narrow range of acidity; therefore, our body uses pH as a first line of defense against infections. Many of our body fluids are overly acidic or alkaline, acting as natural disinfectants. The acidity record goes to the gastric juice (the acidic liquid inside the stomach), which has a pH of 1 to 3, due to the secretion of hydrochloric acid from the glands in its lining. Using its low pH, the stomach eliminates most microorganisms in the food and only a minority of them make it into the intestine. Some bacteria, however, have evolved surprisingly smart systems to thrive even in the most prohibitive conditions of pH. The ulcer-causing bug *Helicobacter pylori*, for example, lives comfortably in the stomachs of millions of people by secreting an enzyme called *urease*, which neutralizes the acid, creating a micro-environment of supportable pH. Skin (and the vaginal mucosa, too) has a slightly acid pH, due to the secretion of lactic acid, which hampers the growth of foreign microorganisms while favoring that of healthy flora. Using neutral soap is a way to preserve the natural pH balance of skin and to keep many unpleasant microbes at bay.

Next time you shower, watch the smooth foam on your body. It's only soap, but it could remind you of the same trick that, in Crichton's fiction, saved humanity from disaster.

SERGIO PISTOI started his career as a molecular biologist. Soon after he finished his Ph.D. in 1994, a radiation incident in his lab turned him into an evil science-writing superhero. He was an intern at *Scientific American* in New York and a stringer for Reuters Health. His credits include *Scientific American*, *New Scientist*, *Nature*, and many Italian print and radio outlets. He is also a consultant for research planning and portfolio management. He is a member of the National Association of Science Writers, NASW, and the European Union of Science Writer's Associations.

He hides in Tuscany, Italy, with a fake identity. He can be found at www.greedybrain.com.

References

Schneider, J. "A Model for Non-chemical Form of Life: Crystalline Physiology." *Origins of Life* (1977): 33–38.

"The Quarantine and Certification of Martian Samples." Committee on Planetary and Lunar Exploration (COMPLEX), Space Studies Board, National Research Council. National Academy Press, 2002. <http://www.nap.edu/catalog/10138.html>

Cairns-Smith, A. G. "The Origin of Life and the Nature of the Primitive Gene." *Journal of Theoretical Biology* 5 (1966): 3–88.

Prusiner, Stanley B. "Detecting Mad Cow Disease." *Scientific American*, July 2004.

Marks, P. "Keeping Alien Invaders at Bay." *The New Scientist*, 28 Apr. 2007.

VIRTUAL REALITY AND MAN-MACHINE INTERFACE IN *DISCLOSURE* AND *THE TERMINAL MAN*

Ray Kurzweil

Two of Michael Crichton's novels, The Terminal Man *and* Disclosure, *address issues concerning the direct interface of machines with the human brain. Today the U.S. military, corporations (Honda being one example), and many academic institutions are experimenting with brain-machine interfaces. Ray Kurzweil answers questions on the feasibility of whether or not you may eventually have a chip implanted in your head, if not one on your shoulder. The answer to this may not surprise you, but the sheer number of chips that may be implanted, what they may do for you, and the time frame in which this actually may occur* will.

The NPS staff has developed a computer that will monitor electrical activity of the brain, and when it sees an attack starting, will transmit a shock to the correct brain area. This computer is about the size of a postage stamp and weighs a tenth of an ounce. It will be implanted beneath the skin of the patient's neck.

—Michael Crichton, *The Terminal Man*

IN MICHAEL CRICHTON'S 1969 thriller *The Terminal Man*, Harold Benson undergoes a surgical procedure to attach his brain to a postage-stamp sized computer. Crichton astutely predicts the potential for implanting a computer chip in the brain to help manage neurological ailments (see "Shock to the System" elsewhere in this volume). Twenty-four years later, in *Disclosure*, Crichton again predicts the future of man-machine interface with his description of a virtual reality device that allows Tom Sanders to literally walk through his company's databases. In both cases, Crichton effectively predicts the direction of future technology and man-machine interface in particular. In both cases, his predictions fall substantially short of the truly exciting developments that are down the road.

I've become a student of technology trends as an outgrowth of my career as an inventor. If you work on creating technologies, you need to anticipate where technology will be at points in the future so that your project will be feasible and useful when it's completed, not just when you started. Over the course of a few decades of anticipating technology, I've become a student of technology trends and have developed mathematical models of how technologies in different areas are developing.

This has given me the ability to invent things that use the materials of the future, and not just limit my ideas to the resources we have today. Alan Kay has noted, "To anticipate the future we need to invent it." So we can invent with future capabilities if we have some idea of what they will be.

Perhaps the most important insight that I've gained, which people are quick to agree with but very slow to really internalize and appreciate all of its implications, is the accelerating pace of technological change itself.

One Nobel laureate recently said to me, "There's no way we're going to see self-replicating nanotechnological entities for at least a hundred years." And yes, that's actually a reasonable estimate of how much work it will take. It'll take a hundred years of progress, at today's rate of progress, to get self-replicating nanotechnological entities. But the rate of progress is not going to remain at today's rate; according to my models, it's doubling every decade. We will make a hundred years of progress at today's rate of progress in twenty-five years. The next ten years will be like twenty, and the following ten years will be like forty. The twenty-first century will therefore be like twenty thousand years of progress—at today's rate. The twentieth century, as revolutionary as it was, did not have a hundred years of progress at today's rate; since we accelerated up to today's rate, it really had about twenty years of progress. The twenty-first century will be about a thousand times greater, in terms of change and paradigm shift, than the twentieth century.

A lot of these trends stem from the implications of Moore's Law. Moore's Law refers to integrated circuits and famously states that the computing power available for a given price will double every twelve to twenty-four months. Moore's Law has become a synonym for the exponential growth of computing. It also explains why Crichton's prediction of a "postage-sized" computer in *The Terminal Man*, daring at the time, now sounds almost ridiculously large.

I've been thinking about Moore's Law and its context for at least twenty years. What is the real nature of this exponential trend? Where does it come from? Is it an example of something deeper and more profound? As I will show, the exponential growth of computing goes substantially beyond Moore's Law. Indeed, exponential growth goes beyond just computation, and applies to every area of information-based technology—technology that will ultimately reshape our world.

Observers have pointed out that the accuracy of Moore's Law is going to come to an end. According to Intel and other industry experts, we'll run out of space on an integrated circuit within fifteen years, because the key features will only be a few atoms in width. So will that be the end of the exponential growth of computing?

That's a very important question to keep in mind as we ponder the nature of the twenty-first century. To address this question, I put forty-nine famous computers on an exponential graph. In the lower left-hand corner is the data processing machinery that was used in the 1890 American census (calculating equipment using punch cards). In 1940, Alan Turing developed a computer based on telephone relays that cracked the German Enigma code and gave Winston Churchill a transcription of nearly all the Nazi messages. Churchill needed to use these transcriptions with great discretion, because he realized that using them could tip off the Germans. If, for example, he had warned Coventry authorities that their city was going to be bombed, the Germans would have seen the preparations and realize that their code had been cracked. However, in the Battle of Britain, the English flyers seemed to magically know where the German flyers were at all times.

In 1952, CBS used a more sophisticated computer based on vacuum

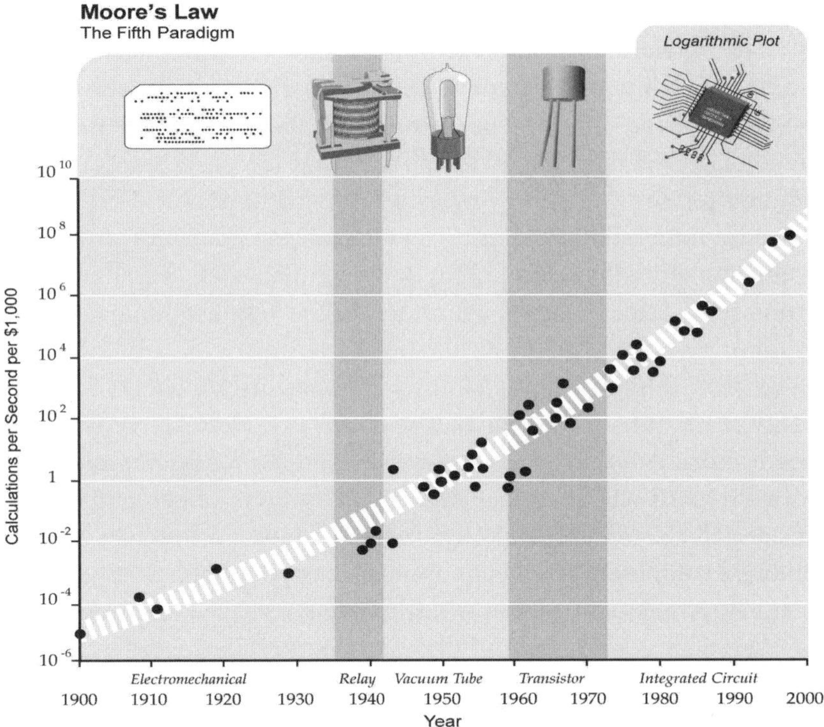

tubes to predict the election of a U.S. president, President Eisenhower. In the upper right-hand corner is the computer sitting on your desk right now.

One insight we can gain from this chart is that Moore's Law was not the first but the fifth paradigm to provide exponential growth of computing power. Each vertical line represents the movement into a different paradigm: electro-mechanical, relay-based, vacuum tubes, transistors, integrated circuits. Every time a paradigm ran out of steam, another paradigm came along and picked up where that paradigm left off.

People are quick to criticize exponential trends, saying that ultimately they'll run out of resources, like rabbits in Australia. But every time one particular paradigm reached its limits, another completely different method would continue the exponential growth. They were making vacuum tubes smaller and smaller but finally got to a point where they couldn't make them any smaller and maintain the vacuum. Then transistors came along, which are not just small vacuum tubes. They're a completely different paradigm.

Every horizontal level on this graph represents a multiplication of computing power by a factor of a hundred. A straight line in an exponential graph means exponential growth. What we see here is that the rate of exponential growth is itself growing exponentially. We doubled the computing power every three years at the beginning of the century, every two years in the middle, and we're now doubling it every year.

It's obvious what the sixth paradigm will be: computing in three dimensions. After all, we live in a three-dimensional world and our brain is organized in three dimensions. The brain uses a very inefficient type of circuitry. Neurons are very large "devices," and they're extremely slow. They use electrochemical signaling that provides only about two hundred calculations per second, but the brain gets its prodigious power from parallel computing resulting from being organized in three dimensions. Three-dimensional computing technologies are beginning to emerge. There's an experimental technology at MIT's Media Lab that has 300 layers of circuitry. In recent years, there have been substantial strides in developing three-dimensional circuits that operate at the molecular level.

Nanotubes, which are my favorite, are hexagonal arrays of carbon atoms that can be organized to form any type of electronic circuit. You can create the equivalent of transistors and other electrical devices. They're physically very strong, with fifty times the strength of steel. The thermal issues appear to be manageable. A one-inch cube of nanotube circuitry would be a million times more powerful than the computing capacity of the human brain.

Over the last several years, the level of confidence in building three-dimensional circuits and achieving at least the hardware capacity to emulate human intelligence has changed. This has raised a more salient issue, namely that "Moore's Law may be true for hardware but it's not true for software." From my own four decades of experience with software development, I believe that is not the case. Software productivity is increasing very rapidly. An example from one of my own companies: in fifteen years we went from a $5,000 speech-recognition system that recognized 1,000 words poorly, without continuous speech, to a $50 product with a 100,000-word vocabulary that's far more accurate. That's typical for software products. With all of the efforts in new software development tools, software productivity has also been growing exponentially, albeit with a smaller exponent than we see in hardware.

Miniaturization is another very important exponential trend. We're making things smaller at a rate of 5.6 per linear dimension per decade. Nanotechnology is not a single unified field only worked on by nanotechnologists. Nanotechnology is simply the inevitable result of a pervasive trend toward making things smaller, which we've been doing for many decades.

Following is a chart of computing's exponential growth, projected into the twenty-first century. Right now, your typical $1,000 PC is somewhere between an insect and a mouse brain. The human brain has about 100 billion neurons, with about 1,000 connections from one neuron to another. These connections operate very slowly, on the order of 200 calculations per second, but 100 billion neurons times 1,000 connects creates 100 trillion-fold parallelism. Multiplying that by 200 calculations per second yields 20 million billion calculations per second, or, in computing terminology, 20 billion MIPS. We'll have 20 billion MIPS for $1,000 by the year 2020.

Now, that won't automatically give us human levels of intelligence, because the organization, the software, the content, and the embedded knowledge are equally important. Below I will address the scenario in which I envision achieving the software of human intelligence. I believe it is clear that we will have the requisite computing power. By 2050, $1,000 of computing will equal one billion human brains. That might be off by a year or two, but the twenty-first century won't be wanting for computational resources.

In *The Terminal Man*, Crichton describes the operation that connects Benson to the computer chip:

> At length he was satisfied. "Are we ready to wire?" he asked the team. They nodded. He stepped up to the patient and said, "Let's go through the dura."
>
> Up to this point in the operation, they had drilled through the skull, but had left intact the membrane of *dura mater* which covered the brain and enclosed the spinal fluid. Ellis's assistant used a probe to puncture the dura.
>
> "I have fluid," he said, and a thin trickle of clear liquid slid down the side of the shaved skull from the hole. A nurse sponged it away (68).

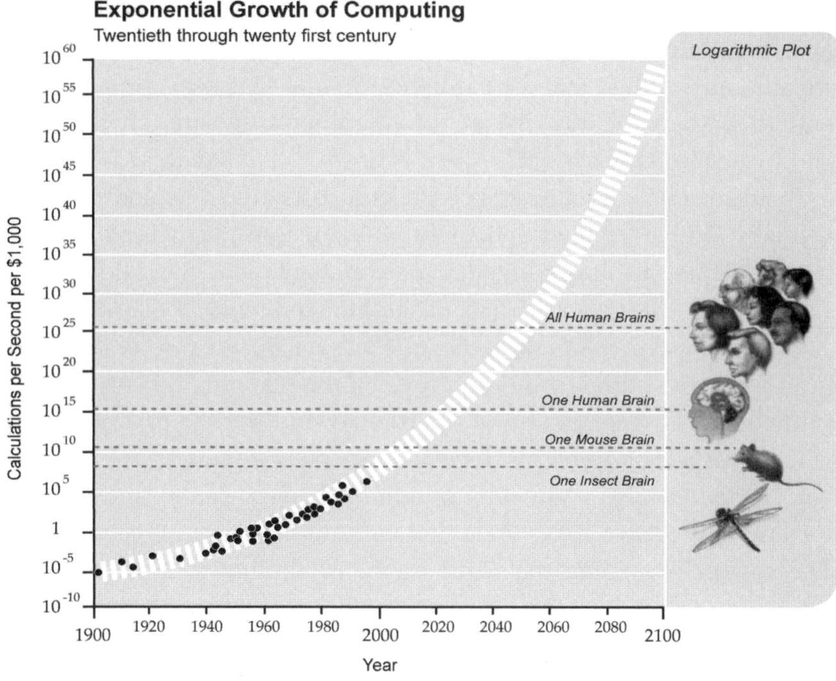

Messy. While we have now reached this level of technology, twenty years from now we'll do quite a bit better. Let's go out to 2029 and put together some of the trends that I've discussed. By that time, we'll be able to build nanobots, microscopic-sized robots that can go inside your capillaries and travel through your brain and scan the brain from inside. We can almost build these kinds of circuits today. We can't make them quite small enough, but we can make them fairly small. The Department of Defense is developing tiny robotic devices called "Smart Dust." The current generation is one millimeter—that's too big for this scenario—but these tiny devices can be dropped from a plane, and find positions with great precision. You can have many thousands of these on a wireless local area network. They can then take visual images, communicate with each other, coordinate, send messages back, act as nearly invisible spies, and accomplish a variety of military objectives.

We are already building blood-cell-sized devices that go inside the bloodstream, and there are four major conferences on the topic of "bioMEMS" (biological Micro Electronic Mechanical Systems). The nanobots I am envisioning for 2029 will not necessarily require their own navigation. They could move involuntarily through the bloodstream and, as they travel by different neural features, communicate with them the same way that we now communicate with different cells within a cell phone system.

Brain-scanning resolution, speeds, and costs are all exploding exponentially. With every new generation of brain scanning we can see with finer and finer resolution. There's a technology today that allows us to view many of the salient details of the human brain. Of course, there's still no full agreement on what those details are, but we can see brain features with very high resolution, provided the scanning tip is right next to the features. We can scan a brain today and see the brain's activity with very fine detail; you just have to move the scanning tip all throughout the brain so that it's in close proximity to every neural feature.

Now, how are we going to do that without making a mess of things? The answer is to send the scanners inside the brain. By design, our capillaries travel by every interneuronal connection, every neuron, and every neural feature. We can send billions of these scanning robots, all on a wireless local area network, and they would all scan the brain from inside and create a very high-resolution map of everything that's going on.

What are we going to do with the massive database of neural information that develops? One thing we will do is reverse-engineer the brain, that is, understand the basic principles of how it works. This is an endeavor we have already started. We already have high-resolution scans of certain areas of the brain. The brain is not one organ; it's comprised of several hundred specialized regions, each organized differently. We have scanned certain areas of the auditory and visual cortex, and have used this information to design more intelligent software. Carver Mead at Caltech, for example, has developed powerful, digitally controlled analog chips that are based on these biologically inspired models from the reverse engineering of portions of the visual and auditory systems. His visual sensing chips are used in high-end digital cameras.

We have demonstrated that we are able to understand these algorithms, but they're different from the algorithms that we typically run on our computers. They're not sequential and they're not logical;

they're chaotic, highly parallel, and self-organizing. They have a holographic nature in that there's no chief-executive-officer neuron. You can eliminate any of the neurons, cut any of the wires, and it makes little difference—the information and the processes are distributed throughout a complex region.

Based on these insights, we have developed a number of biologically inspired models today. This is the field I work in, using techniques such as evolutionary "genetic algorithms" and "neural nets," which use biologically inspired models. Today's neural nets are mathematically simplified, but as we get a more powerful understanding of the principles of operation of different brain regions, we will be in a position to develop much more powerful, biologically inspired models. Ultimately we can create and recreate these processes, retaining their inherently massively parallel, digitally controlled analog, chaotic, and self-organizing properties. We will be able to recreate the types of processes that occur in the hundreds of different brain regions, and create entities—they actually won't be in silicon, they'll probably be using something like nanotubes—that have the complexity, richness, and depth of human intelligence.

Understanding the methods of the human brain will help us to design similar biologically inspired machines. Another important application will be to actually interface our brains with computers, which I believe will become an increasingly intimate merger in the decades ahead, far beyond what Crichton imagined over thirty years ago.

Miguel Nicolelis and his colleagues at Duke University have implanted sensors in the brains of monkeys, enabling the animals to control a robot through thought alone. The first step in the experiment involved teaching the monkeys to control a cursor on a screen with a joystick. The scientists collected a pattern of signals from EEGs (brain sensors) and subsequently caused the cursor to respond to the appropriate patterns rather than physical movements of the joystick. The monkeys quickly learned that the joystick was no longer operative and that they could control the cursor just by thinking. This "thought detection" system was then hooked up to a robot, and the monkeys were able to learn how to control the robot's movements with their thoughts alone. By getting visual feedback on the robot's performance, the monkeys were able to perfect their thought control over the robot. The goal of this research is to provide a similar system for paralyzed humans that will enable them to control their limbs and environment.

A key challenge in connecting neural implants to biological neurons is that the neurons generate glial cells, which surround a "foreign" object in an attempt to protect the brain. Ted Berger and his colleagues are developing special coatings that will appear to be biological and therefore attract rather than repel nearby neurons.

Another approach being pursued by the Max Planck Institute for Human Cognitive and Brain Sciences in Munich is directly interfacing nerves and electronic devices. A chip created by Infineon allows neurons to grow on a special substrate that provides direct contact between nerves and electronic sensors and stimulators. Similar work on a "neurochip" at Caltech has demonstrated two-way, noninvasive communication between neurons and electronics.

Our machines today are still a million times simpler than the human brain, which is one key reason that they still don't have the endearing qualities of people. They don't yet have our ability to get the joke, to be funny, to understand people, to respond appropriately to emotion, or to have spiritual experiences. These are not side effects of human intelligence, or distractions; they are the cutting edge of human intelligence. It will require a technology with the complexity of the human brain to create entities that have those kinds of attractive and convincing features.

> "The scanner. Mapping you. Don't worry about it. Here's the headset." He brought the headset down from the ceiling and started to place it over her eyes.
>
> "Just a minute." She pulled away. "What is this?"
>
> "The headset has two small display screens. They project images right in front of your eyes. Put it on. And be careful. These things are expensive."
>
> "How expensive?"
>
> "A quarter of a million dollars apiece." He fitted the headset over her eyes and put the headphones over her ears (*Disclosure* 431–432).

In *Disclosure*, Crichton envisions a large virtual reality machine which includes not only large headsets with two screens inside and trampoline-sized walker pads but also external scanners and computer. This may be a realistic conception using today's technology, but looking ahead it's quite cumbersome. Consider a scenario involving a direct connection between the human brain and these nanobot-based implants. There are a number of different technologies that have already been demonstrated for communicating in both directions between the wet, analog world of neurons and the digital world of electronics. One such technology, called a neurotransistor, provides this two-way communication. If a neuron fires, this neuron transistor detects that electromagnetic pulse; it's communication from the neuron to the electronics. It can also cause the neuron to fire or prevent it from firing.

For full-immersion virtual reality, we will send billions of these nanobots to take up positions by every nerve fiber coming from all of our senses. If you want to be in real reality, they sit there and do nothing. If you want to be in virtual reality, they suppress the signals coming from our real senses and replace them with the signals that you would have been receiving if you were in the virtual environment.

In this scenario, we will have virtual reality from within and it will be able to recreate all of our senses. These will be shared environments, so you can go there with one person or many people. Going to a Web site will mean entering a virtual reality environment encompassing all of our senses, and not just the five senses, but also emotions, sexual pleasure, humor. There are actually neurological correlates of all of these sensations and emotions, which I discuss in my book *The Age of Spiritual Machines*.

For example, surgeons conducting open-brain surgery on a young woman (while awake) found that stimulating a particular spot in the girl's brain would cause her to laugh. The surgeons thought that they were just stimulating an involuntary laugh reflex. But they discovered that they were stimulating the perception of humor: whenever they stimulated this spot, she found everything hilarious. "You guys are just so funny standing there" was a typical remark.

Using these nanobot-based implants, you will be able to enhance or modify your emotional responses to different experiences. That can be part of the overlay of these virtual reality environments. You will also be able to have different bodies for different experiences. Just as people today project their images from Web cams in their apartment, people will beam their whole flow of sensory and even emotional experiences out on the Web, so you can, à la the plot concept of the movie *Being John Malkovich*, experience the lives of other people.

In virtual reality we won't be restricted to a single personality, since we will be able to change our appearance and effectively become other people. Without altering our physical body (in real reality) we will be able to readily transform our projected body in these three-dimensional virtual environments. We can select different bodies at the same time for different people. So your parents may see you as one person, while your girlfriend will experience another. However, the other person may choose to override your selections, preferring to see you differently than the body you have chosen for yourself. You could pick different body projections for different people: Ben Franklin for a wise uncle, a clown for an annoying coworker. Romantic couples can choose whom they wish to be, even to become each other. These are all easily changeable decisions. These possibilities go far beyond *Disclosure*'s crude virtual reality machine, which envisions the image of Tom Sander's body as merely recorded by video and then projected onto the screens inside his headset.

I had the opportunity to experience what it is like to project myself as another persona in a virtual reality demonstration at the 2001 TED (technology, entertainment, design) conference in Monterey. By means of magnetic sensors in my clothing a computer was able to track all of my movements. With ultra-high-speed animation the computer created a life-size, near photorealistic image of a young woman—Ramona—who followed my movements in real time. Using signal-processing technology, my voice was transformed into a woman's voice and also controlled the movements of Ramona's lips. So it appeared to the TED audience as if Ramona herself were giving the presentation.

The experience was a profound and moving one for me. When I looked in the "cybermirror" (a display showing me what the audience was seeing), I saw myself as Ramona rather than the person I usually see in the mirror. I experienced the emotional force—and not just the intellectual idea—of transforming myself into someone else.

Ultimately, these nanobots will expand human intelligence and our abilities and facilities in many different ways. Because they're communicating with each other wirelessly, they can create new neural connections. These can expand our memory, cognitive faculties, and pattern-recognition abilities. We will expand human intelligence by expanding its current paradigm of massive interneuronal connections as well as through intimate connection to non-biological forms of intelligence.

We will also be able to download knowledge, something that machines can do today that we are unable to do. For example, we spent several years training one research computer to understand human speech using the biologically inspired models—neural nets, Markov models, genetic algorithms, self-organizing patterns—that are based on our crude current understanding of self-organizing systems in the biological world. A major part of the engineering project was collecting thousands of hours of speech from different speakers in different dialects and then exposing this to the system and having it try to recognize the speech. It made mistakes, and then we had it adjust automatically, and self-organize to better reflect what it had learned.

Over many months of this kind of training, it made substantial improvements in its ability to recognize speech. Today, if you want your personal computer to recognize human speech, you don't have to spend years training it the same painstaking way, as we need to do with every human child. You can just load the evolved models—it's called "loading the software." So machines can share their knowledge.

We don't have quick downloading ports on our brains. But as we build non-biological analogs of our neurons, interconnections, and neurotransmitter levels where our skills and memories are stored, we won't leave out the equivalent of downloading ports.

When you talk to somebody in the year 2040, you will be talking to someone who may happen to be of biological origin but whose mental processes are a hybrid of biological and electronic thinking processes, working intimately together. Instead of being restricted, as we are today, to a mere hundred trillion connections in our brain, we'll be able to expand substantially beyond this level. Our biological thinking is flat; the human race has an estimated 10^{26} calculations per second, and that biologically determined figure is not going to grow. But non-biological intelligence is growing exponentially. The crossover point, according to my calculations, is in the 2030s; some people call this the Singularity.

The Singularity will represent the culmination of the merger of our biological thinking and existence with technology, resulting in a world that is still human but that transcends our biological roots. There will be no distinction, post-Singularity, between human and machine or between physical and virtual reality. If you wonder what will remain unequivocally human in such a world, it's simply this quality: ours is the species that inherently seeks to extend its physical and mental reach beyond current limitations.

As we get to 2050, the bulk of our thinking—which in my opinion is still an expression of human civilization—will be non-biological. The non-biological portion of our thinking will still be human thinking, because it's going to be derived from human thinking. It will be created by humans, or created by machines that are created by humans, or created by machines that are based on reverse engineering of the human brain or downloads of human thinking, or one of many other intimate connections between human and machine thinking that we can't even contemplate today.

The range of intelligent tasks in which machines can now compete with human intelligence is continually expanding. In a cartoon I designed for *The Age of Spiritual Machines*, a defensive "human race" is seen writing out signs that state what only people (and not machines) can do. Littered on the floor are the signs the human race has already discarded because machines can now perform these functions: diagnose an electrocardiogram, compose in the style of Bach, recognize faces, guide a missile, play Ping-Pong, play master chess, pick stocks, improvise jazz, prove important theorems, and understand continuous speech. Back in 1999 these tasks were no longer solely the province of human intelligence; machines could do them all.

On the wall behind the man symbolizing the human race are signs he has written out describing the tasks that were still the sole province of humans: have common sense, review a movie, hold press conferences, translate speech, clean a house, and drive cars. If we were to redesign this cartoon in a few years, some of these signs would also be likely to end up on the floor.

A common reaction to this is that it is a dystopian vision, because I am "placing humanity with the machines." This same dystopian fear of technology is manifested in Harold Benson's eventual malfunction when his violent seizures become uncontrollable due to the programming of his computer. But that's because most people have a prejudice against machines. Most observers don't truly understand what machines are ultimately capable of, because all the machines that they've ever "met" are very limited, compared to people. But that won't be true of machines circa 2030 and 2040. When machines are derived from human intelligence and are a million times more capable, we'll have a different respect for machines, and there won't be a clear distinction between human and machine intelligence. Although the Singularity has many faces, its most important implication is this: our technology will match and then vastly exceed the refinement and suppleness of what we regard as the best of human traits. We will effectively merge with our technology.

We are already well down this road. If all the machines in the world stopped today, our civilization would grind to a halt. That wasn't true as recently as thirty years ago. In 2040, human and machine intelligence will be deeply and intimately melded. We will become capable of far more profound experiences of many diverse kinds. We'll be able to "recreate the world" according to our imaginations, with implications far more profound than anything predicted in *Disclosure*, and far less dystopian than predicted in *The Terminal Man*.

RAY KURZWEIL was the principal developer of the first CCD flatbed scanner, the first omni-font optical character recognition, the first print-to-speech reading machine for the blind, the first text-to-speech synthesizer, the first music synthesizer capable of recreating the grand piano and other orchestral instruments, and the first commercially marketed large-vocabulary speech recognition. Ray's Web site KurzweilAI.net has more than 1 million readers. Among Ray's many honors, he is the recipient of the $500,000 MIT-Lemelson Prize, the National Medal of Technology, and the National Inventor's Hall of Fame, established by the U.S. Patent Office. He has received fifteen honorary doctorates and honors from three U.S. presidents.

Ray has written five books, four of which have been national bestsellers. *The Age of Spiritual Machines* has been translated into nine languages and was the #1 bestselling book on Amazon in science. Ray's latest book, *The Singularity Is Near*, was a *New York Times* bestseller, and has been the #1 book on Amazon in both science and philosophy.

References

Crichton, Michael. *The Andromeda Strain/The Terminal Man*. Harper: 2002.

Crichton, Michael. *Disclosure*. Random House, 1993.

Kurzweil, Ray. *The Age of Spiritual Machines: When Computers Exceed Intelligence*. New York: Penguin, 2000.

Kurzweil, Ray. *The Singularity Is Near.* New York: Penguin, 2005.

SHOCK TO THE SYSTEM

Steven Gulie

When Michael Crichton wrote The Terminal Man, *he did so after seeing a patient in a hospital with electrodes implanted in his brain. Believing that the public had little or no knowledge of such procedures, he wrote the book in part to call attention to the practice. Although the particular method of treatment Crichton witnessed back then is no longer practiced today, Steven Gulie has taken it as his charge to relate his stimulating firsthand experience with a similar procedure—one that was intentionally a shock to his system.*

I'M LYING IN AN OPERATING ROOM at the Stanford University hospital, head shaved, waiting for my brain surgery to begin. Sure, I feel anxious, but mostly I feel crowded. There are ten people milling about, tinkering with instruments and examining me. It's an impressive crew: a neurosurgeon and his fellow, a neurologist and her fellow, an anesthesiologist, an experimental physicist, and a graduate student in electrical engineering. That's right, a physicist and an electrical engineer. Directly behind me, out of my sight, is the star of the show, chief neurosurgeon Jaimie Henderson: forty-four years old, tall, erudite, and handsome. On my right, flexing my hands, is neurologist Helen Bronte-Stewart: brisk, smart, and beautiful. In fact, nearly everyone here is not only brilliant but also pretty enough to play themselves in the movie version of this story. I call them Team Hubris.

Today I'm an honorary member. I'll be kept awake for the entire procedure. During the surgery I will talk and move my limbs on command, which helps Team Hubris know which part of my brain is being poked.

Unfortunately, this also means I'm conscious when Henderson produces what looks like a hand drill and uses it to burr two dime-sized holes into the top of my skull. It doesn't hurt, but it's loud.

Team Hubris is installing a deep brain stimulator, essentially a neurological pacemaker, in my head. This involves threading two sets of stiff wires in through my scalp, through my cerebrum—most of my brain—and into my subthalamic nucleus, a target the size of a lima bean, located near the brain stem. Each wire is a little thinner than a small, unfolded paper clip, with four electrodes at one end. The electrodes will eventually deliver small shocks to my STN. How did I get into this mess? Well, I have Parkinson's disease. If the surgery works, these wires will continually stimulate my brain in an attempt to relieve my symptoms.

The first half of the operation goes smoothly. Through amplifiers connected to the probes, the team can listen to my neurons fire as the wires move through my cortex. This helps pinpoint the location of the probes. When I move a limb, for example, my subthalamic nucleus ignites, resulting in a loud burst of static. The doctors and the physicist practically sing along: "That's driving it!" "Yes!" "Listen to that!" "Dorsiflexion!" "Perfect!" The fellows' eyes are shining. Everyone looks happy and expectant—proud. Following instructions from the neurologist, I tap my fingers, open and close my mouth, stick out my tongue. She is pleased.

After positioning the first electrode, which takes about three hours, we stop for ten minutes to prep for the second electrode. I listen to my iPod; several doctors step out to stretch. For some reason, when we're ready to start again, the neurologist, Bronte-Stewart, isn't there. They page her. They wait. They page her again. (She later says she never got the first page). Eventually, they proceed without her. The neurology fellow steps in, flexing my hands and asking me to stick out my tongue. She doesn't seem too happy.

She's not the only one. Team Hubris is struggling with the second electrode. "That sounds like it, but it's dorsal, too dorsal." "I don't like this electrode." "I don't think it's the electrode." "I think maybe it's a software problem." "Try another cable." "Another audio cable?" "Well. . . yes."

Bronte-Stewart returns at last. She relieves the fellow and begins flapping my wrist and flexing my legs. She seems upset. Meanwhile, the physicist and the engineer continue working on the errant electrode. This is not good.

The tremors started in 1999. I remember pouring a glass of wine and my hand starting to shake. "What's that?" my wife at the time asked. "Parkinson's," I joked. We laughed. I was only forty-three. It seemed funny.

The shaking went away, but over the next few weeks I started using the computer mouse and eating with my left hand. I didn't know it, but the Parkinson's was affecting the fine motor control on my right side. It was subtle, but enough to make me more comfortable as a southpaw. "That's odd," I thought. "When did I become left-handed?"

Then I quit surfing. As a Santa Cruz, California, local, I had gone out on the water at Sewer Peak or Steamer Lane almost every day for years. Suddenly I wasn't catching any waves. I just couldn't get the nose of my board down, couldn't coordinate the movements to shift my weight forward and transfer the swell's momentum to my board. "I'm getting old," I thought. "Old and fat. And weak."

But I was getting old too fast. I felt like I was seventy. Finally, in 2000, I went to the doctor and got the official diagnosis: Parkinson's disease. It affects about one person in 100, usually people in their seventies. I was in my forties. That made me more like one in 4,000. That's like winning the lottery. Whoops, wrong lottery.

Parkinson's causes brain cells in the substantia nigra (Latin for "black stuff") to die off. This area is where the neurotransmitter dopamine gets manufactured, and without dopamine, the brain's circuits start misbehaving. When the substantia nigra is 50 to 80 percent gone, you begin to experience Parkinson's symptoms: usually tremors but also constipation, stiffness, and depression. You slowly lose smooth motion of all kinds, as well as the ability to feel pleasure. It's a darkness that covers you.

Doctors can't give you replacement dopamine to fix the problem, because dopamine can't cross the blood-brain barrier—the fine mesh that keeps coarse intruders like bacteria out of the brain. Instead they give you levodopa (L-dopa), a dopamine precursor with molecules small enough to slip through the barrier. It's like flooding an assembly line with partly manufactured goods; if one of the workers is in the mood to make a little dopamine, it's easy to finish the product.

But many of the workers on this particular assembly line are already dead, and the rest aren't feeling so good. Consequently, production is spotty. And there are side effects. Most of the uncontrolled movements we associate with Parkinson's aren't actually symptoms of the disease; they're caused by L-dopa.

As time goes on and you need more L-dopa to produce any dopamine at all, the side effects become worse. It gets so you can't walk, your movements are so out of control. But it's still better than the alternative: no dopamine, leading to paralysis, the inability to swallow, and death.

Now there's another alternative: a personal brain implant. Doctors can use scans to detect aberrant electrical activity in the brain; they can even localize it in three dimensions through computed tomography (CT scanning). Using this information, they can place electric stimulators directly on a trouble spot. In the case of Parkinson's, the stimulators zap either the globus pallidus interna or the subthalamic nucleus, depending on the type of symptoms you have. The shocks seem to let these parts of the brain function normally despite the lack of dopamine. Over the last decade, the procedure has been performed on more than 20,000 Parkinson's patients.

Combating Parkinson's with a flood of drugs is like carpet bombing. It's better to zap the area in a surgical strike—it's more effective, and there's far less collateral damage. Researchers are now experimenting with brain stimulators for epilepsy, obsessive-compulsive disorder, Tourette's syndrome, and depression. Animal studies are under way to use brain stimulators to treat eating disorders. It's a whole new wave of personal digital technology.

Still, when it's you, the idea of having someone saw open your skull and insert wires into your brain doesn't seem like a very good idea. In fact, it seems like a really bad idea. But the Stanford team is one of the most experienced in the world. They are fairly confident that a deep brain stimulator will greatly reduce my Parkinson's symptoms; that it will turn the clock back a year, two years, even five; that I will need to take less L-dopa, have fewer side effects, and live a much more normal life.

By my first meeting with Team Hubris, my life hadn't been normal for a while: I was already stooped and shuffling, dropping things, complaining about my infrequent bowel movements. I couldn't ride a bike. In fact, I had trouble tying my shoes. I could barely type. And the side effects of L-dopa were already showing: flailing motions, weird posturing. Eventually I was forced to take disability leave from my job as a technical writer at Apple. I felt as though I was slowly fading out of other people's lives, out of life itself. Wires in the brain? Sign me up.

Jaimie Henderson was tinkering with the brains of Parkinson's patients back when it wasn't fashionable—in the early '90s, when instead of stimulating the brain, doctors would destroy the troublesome parts of it altogether. This was the type of operation Michael J. Fox had in 1998, when doctors surgically eradicated a section of his thalamus. It was effective, but crude. Call it brain surgery 1.0.

In 1995, doctors at Mt. Sinai Medical Center in New York conducted the first U.S. surgery to put stimulators directly into the subthalamic nucleus. Henderson started performing the procedure experimentally in 1999, and in 2002 the FDA approved the use of brain stimulators for Parkinson's. Good health care plans—like the one offered by my employer, Apple—cover the procedure. The cost for mine: about $250,000.

The heart of the device is a small computer that is implanted under the collarbone. Electricity flows from this device—the stimulator—through wires running under the skin and the scalp, through the electrodes into the brain, and returns to the computer through the body to close the circuit. The power is always on, so the stimulation is continual. The apparatus is battery-operated, and the battery is not rechargeable. They have to do minor surgery to swap it out every three to five years.

The system can be fine-tuned after the operation by activating various electrodes, shifting the affected area by a millimeter or two. The doctors can also tweak the frequency and amplitude of the electrical stimulation, modify the pulse width, and make other adjustments to the software through a remote control. Wireless? Software? Now that's brain surgery 2.0.

I asked Henderson about features still on the whiteboard: What will be in rev 2.2 or 2.5? He thinks the next release of the stimulator will sense chaotic activity in the brain and turn itself on only when needed. That's on par with current heart pacemakers, which no longer mindlessly zap you with a steady pulse but actually look for a problem to fix. The next-gen device will also probably be transdermally rechargeable, so you won't need surgery to get new batteries.

It's tempting to wait. But as with any tech product, there will always be the promised next release full of new features. Besides, as Henderson stresses, the current model is a "stable release." Right. I've got enough problems without having to debug my brain implant.

There will likely be side effects. I may experience speech problems or difficulty finding words. The doctors will try to minimize that by placing the electrodes just right, but things may never be exactly the same. This leaves me with a lingering question: To the extent that I'm a person with Parkinson's, no, I won't be me anymore; as for the rest, we'll have to wait and see? Well, it depends on how you define "me," doesn't it? In the sense that "me" is a person weighed down with Parkinson's and almost unable to type or tie my shoes, no. I won't be me anymore. As for the rest, we'll see.

The surgery normally goes like this: You get one side of your head wired—this takes between three and four hours. Then you wait a week and they do the other side. Wait a week more and get the pacemaker inserted. Wait till the swelling goes down—maybe another couple of weeks—and get it programmed.

But I'm young and strong, and Team Hubris has a hard time coordinating all these surgeries. So they decide to wire both sides of my head in one session, over roughly six hours, to simplify things. The day before the operation, I have screws inserted in my skull. Yeah, screws.

In traditional brain surgery, your head is bolted into a rigid metal frame while your brain is scanned to provide a 3-D model to work from. This model helps doctors plan a pathway to a precise spot in the brain, avoiding major vessels and arteries (nick one of these and it's game over). If your head moves, it no longer corresponds to the model on the screen. This approach is uncomfortable, and it puts the surgeons under serious time pressure.

The screws are an innovation pioneered by Henderson. They allow surgeons to work without a frame for your head. The screws are driven right into the bone using a cordless screwdriver with a Phillips-head bit. It hurts, but not as much as you might expect. When you get scanned for the 3-D brain model, the screws show up on the scan, creating stable reference points to work from, much the way GPS satellites work. The team can triangulate any spot in the brain using three screws. Four is better, to prevent ambiguity, and five—the number I get—is belt-and-suspenders solid. During the surgery, you can move and talk, because when your head moves, the reference screws move with it. A small robotic insertion device is clamped directly to your head with more screws, effectively turning your skull into the support frame.

This approach also means that after the initial scan, I get to go home, giving the surgeons the afternoon and evening to plan my operation. Instead of being locked into a metal brace, I sit on the couch sipping Chardonnay and eating Vicodin—five titanium bolts jutting from my skull.

Back in surgery, things aren't going as planned. Lying on the table, I'm starting to get very worried. The second electrode still doesn't sound right.

Then something wonderful happens. It's hard to describe, but for more than five years my right hand hasn't felt the way it should. Suddenly, it's back. I can tap my fingers, move freely. It's the miracle cure for Parkinson's I've been reading about! I tell the neurologist.

She seems unconvinced. I'm saying they've hit the sweet spot, but it may be that they've hit the wrong sweet spot. There are structures near the subthalamic nucleus that affect mood, and the doctors don't want to place an electrode there. They aren't trying to make me happy, like some lab rat with an electrode implanted in its pleasure center; they're trying to cure my Parkinson's. "You feel euphoric?" she asks.

"No, no," I say. "It's just that my hand, my hand is back. It's been years since it felt right."

"And this makes you feel how? Happy?"

The rest of the team starts making grumpy noises. The neurosurgeon calls for another electrode, but the physicist assures him that's not the problem. I feel tired and worried. I mention this to the neurologist.

"He has feelings of imminent doom." Well, I don't know about imminent doom, but. . . .

The neurosurgeon tries something. It hurts like a sonofabitch. Ow. Ow! I have no idea what's going on; I didn't think it could hurt—there are no pain receptors in the brain. Is this a stroke? Am I dying?

The doctors decide to stop the surgery. They staple me shut and cart me down to the CT scanner. This must be a stroke, I think, one of the chief dangers of deep brain stimulator surgery. My last moments will be in this Stanford hospital room, looking at a stain on the ceiling above the GE logo on the scanner.

But no. After an agonizing twenty-minute wait, the anesthesiologist and the nurse return, kindly looks on their faces. Fine. Everything is fine. No bleeding. No problems. It most likely hurt because the local anesthetic wore off, and the surgeon touched the margins of the scalp wound while trying to position the probe. The issues were actually relatively minor, they assure me, and well within the bounds of normal operating procedure.

Hiccups happen. In this case, the doctors suspect that the brain shifted a millimeter or so. Things weren't quite where the model said they should be. This can result from either loss of cerebrospinal fluid or simple agitation. It's why they usually do the two sides of the brain a week apart, with fresh scans each time. They can finish the other side later, they assure me, no problem. Next week, or the week after.

In the days after the surgery, my Parkinson's symptoms are remarkably diminished. This is called the microlesion effect. Apparently just the swelling from the poking around is enough to make things better for a while. It fades, but it's awfully encouraging. For about five years now, I've been living without hope. This is a nice change.

No one really knows precisely why deep brain stimulation works. Some things about the deep brain structures, like the thalamus, are understood well enough for stimulators to be routinely successful. But the high-level brain structures in the neocortex, where all the evolutionary action has been for the past 100,000 years or so, are still largely mysteries. How does shocking the thalamus in the deep brain help the cortex in the upper-level brain control fine motor movement? Is this suppressing electrical activity or enhancing it?

For the second surgery, I agree to do my part for science, volunteering for a battery of brain tests while they have my hood up. Before the procedure, Henderson shows me a little gold grid, about half the size of my fingernail. He will lay this grid on my cortex, and it will register neurons firing as the doctors have me perform simple exercises. "It has 100 wires," he says proudly.

I try to look impressed, but I'm thinking, "Only a hundred wires?" To be fair, the hundred wires are actually 100 silicon microprobes (each 0.06 inch long) packed into a 0.16- by 0.16-inch grid. When the grid is implanted into the cortex, each microprobe records the activity from at least one neuron, and sometimes as many as three or four. Right now this is the maximum amount of information we can extract from the human brain.

It's neat-looking and compact, but my first Apple II computer had 1,000 transistors in that space, not 100 wires. It wasn't long before my PC had a processor with 100 million transistors. Just 100 wires? It's a reminder that this technology is still in its infancy.

After the experiments, the second surgery goes about as smoothly as possible. No sooner do I stick out my tongue and tap my fingers a few times, it seems, than it's done. In record time.

They test the electrode placement by putting a little voltage through the wires. There's a ferocious buzzing, like a swarm of bees in my head. They try a few modulations, and the buzzing goes away.

Finally, the anesthesiologist cranks up the gas, and I'm off to never-never land while the surgeons runs the wires under my scalp, places the pacemaker under my collarbone, and closes me up. They had planned to do the pacemaker later, but things have gone so quickly that they do it now.

I wake up as they wheel me into recovery, which at Stanford is a kind of fun house sideshow. People in various states of undress—many of us having just had parts removed or new parts installed—loll and roll about in pain and confusion, all under the watchful eyes of a room full of nurses, orderlies, and aides. The occasional doctor breezes through to provide expert advice or—because this is a teaching hospital—comic relief. The nurses, rolling their eyes, patiently guide the young doctors like sergeants working with newly minted lieutenants.

The pacemaker itself hurts more than I expected. It feels—and looks—like I've been stabbed in the chest. There's clearly no room under my collarbone for this thing, and it bulges out like a first-generation iPod in a tight shirt pocket. The wires aren't the little hair-thin fibers that I expected, either. They're as thick as speaker wires. Which makes sense, because they've got to take a lot of wear and tear without breaking. But the whole thing is more obtrusive than I realized it was going to be. And more painful.

A month after the second surgery, I'm back at Stanford to program the stimulator. Getting the settings right is midway between an art and a science. On each side of the brain is a probe with four electrodes. The team needs to decide which electrodes to activate with how much voltage. The device is capable of delivering 10.5 volts, but at that power there's danger of damaging the brain tissue. So we start at 2 volts and won't go over 3.5.

Raising or lowering the voltage changes the size of the area being stimulated. If the electrode is too near a structure like the internal capsule, the stimulation can cause muscle contractions; too near the substantia nigra, it causes hypomania or depression. Turning the voltage down reduces the affected area so it doesn't cause the side effects, but it makes the whole device less effective.

If necessary, doctors can activate two electrodes so the current flows from one to the other, rather than from the electrode back to the pacemaker in my chest. This drastically shrinks the affected area.

Ultimately, getting the system to work right comes down to trial and error. There are 1,200 possible settings, and fatigue alone prevents testing more than a few at a time. The team starts by trying each electrode on each side. Then they click up the voltage until my tongue sticks to the roof of my mouth, back it down until I feel nothing worse than a slight tingling, all while testing me for Parkinson's symptoms by having me tap my fingers and twirl my wrists as if I were beating eggs.

We hit the sweet spot on both sides at 2.5 volts. I can tap my fingers and scramble eggs like gangbusters, with no side effects. I take a Parkinson's test and ace it. I have no observable signs of Parkinson's except for a tendency to fatigue rapidly. I walk out of the hospital, click my heels in the air, pick up my new fiancée, and swing her around. It's the happiest day of my life.

Over the next twenty-four hours, my symptoms return. Henderson tells me to try increasing the voltage by using the remote control he gave me and to add some Parkinson's medication to the mix. I'm able to get good symptomatic relief, but there are other problems. For starters, I have to turn the device way down to sleep. And I can't tell a joke—my timing is off. My natural gift for mimicry is also gone, as it seems I've lost some of the fine control over my vocal cords. I'm hesitant in social settings: By the time I can muster a response, the topic of conversation has moved on. I'm slightly out of phase with everyone I talk to. And I can't write worth a damn.

At first, the neurology team has a hard time zeroing in on the problem. They can't test for the impairments I'm experiencing, and since none of them can mimic an accent or tell a joke properly either, they don't have much to go on.

Finally, after three months of tinkering, I find another neurologist, Eric Collins, who gets it. He has me count backward from 100 by sevens. With the device off, no problem. With it on, I can't do it. We change the settings until I can. We have to go to two active electrodes on the right side instead of one. He has me recite poetry from memory and fine-tunes me again. Better, almost there, but I'm too tired to continue. And I still can't write. It's like being in a fog.

I e-mail Henderson, describing the problems and the changes the new neurologist has made, and he suggests reversing the polarity on the right side. He knows what he's doing and it helps—a lot. I request that they go to two electrodes on the left side as well. Bingo. After these changes, my head clears. The fog begins to lift. The last piece of the puzzle is adding a new drug, Exelon, just approved for the cognitive problems associated with Parkinson's. That does it. I can finally think and write again.

Today, eight years since the first signs of Parkinson's and after months of fiddling, my body is almost free of symptoms. With the stimulator turned off, a Parkinson's test shows twenty significant impairments. With the stimulator on, it drops to two. Add just a touch of L-dopa and it drops to zero.

The last wisps of fog have cleared away. My jokes make people laugh again. I can keep up with conversation. I can ride a bike. I can write. It's been five months since the surgery, but it has finally all come together: It works. I forget that I even have Parkinson's most of the time. And last November, I went back to work full-time. It's a miracle. A second chance at life.

I know it's not a cure. Parkinson's is degenerative. Those neurons in the brain keep dying, producing less and less dopamine. How long will I feel normal? No one knows. A long-term study completed in 2004 showed that four years after surgery, patients still typically required 50 percent less L-dopa than they did before.

After that, we'll see. The surgery has only had FDA approval since 2002. The long-term effects are simply unknown—I'm the guinea pig. The trick now is to make the most of the time I've been given.

Hand me that bar of surf wax, will you? I haven't taken this board out in a long time, and the Internet is forecasting six- to eight-foot swells, with clear skies.

STEVEN GULIE is a writer, poet, and photographer. He lives in Oakland, California, with his angelic wife and two enchanting daughters, two and a half cats, tanks full of fish, a couple of snakes, and some crickets and preying mantises. Some of the animals eat each other; He tries not to worry about it. He works for Apple, where his job is to take very complex things and explain them simply. He still rides his bike and occasionally body surfs.

NEANDERTHALS AND WENDOLS

Ian Tattersall

Michael Crichton has openly admitted that he wrote Eaters of the Dead *on a bet that he could make* Beowulf *an entertaining story. How did he do that? For starters, by turning the evil Grendel into the Wendol, the last tribe of Neanderthals. Did he succeed? Ian Tattersall elaborates on whether or not a tribe of Neanderthals make a suitable antagonist for the story's hero, Ibn Fadlan.*

EANDERTHALS. If they didn't exist, we'd have to invent them. Wherever on the planet they come from, human beings seem to hunger for a powerful image to embody their darker side, their shadowy connection to the animal world of which we all so clearly form a part but from which we equally strongly feel distinct. Such images come in any number of forms; but in our modern culture the safely extinct but well-documented Neanderthals fit the bill perfectly, although they certainly don't have a monopoly—witness the ongoing dramas of Bigfoot in the Pacific Northwest, or the Almas in the Altai Mountains of western Asia, or the Yeti in the Himalayas. Yet is it just possible that all of these various alter egos are actually the same thing? If they are (or even, for that matter, if they are not), do they correspond to a real but elusive biological entity that is living out there, somewhere? Or are they nothing more (or less) than an expression of some widespread and deep-seated psychic need that resides only within the human skull?

Far be it from me to spoil the fun and claim that there is no human-like form lurking out there in the backwoods or among the snowy wastes. All I am in a position to do is to point out that all assertions to this effect so far have been based on evidence that is inconclusive at best and that has on occasion proved to be out-and-out fraudulent. Still, claims of this kind never seem to go away; and, until some intrepid explorer brings a frightful half-human/half-ape chimera home, might there be any grounds for believing that what we have in such legends is the emanation of a lingering folk memory of some kind? A psychic remnant of a time when humankind was not alone in the world? After all, a pretty good fossil record shows that throughout the seven-million-year history of our hominid family (that is to say, us, plus all our extinct relatives who were not more closely related to the apes), there has typically been more than one kind of hominid alive on Earth. Indeed, there is good evidence that, at some points in hominid history at least, several different kinds of hominid coexisted on the very same landscapes. So, although being alone in the world is what we know and take for granted as normal, it is in fact a rather atypical situation.

Ponderings of this kind rather naturally bring us to the question of who, or what, the beings might have been that, according to Michael Crichton's *Eaters of the Dead*, the Arab traveler Ibn Fadlan encountered in the wilds of tenth-century Scandinavia. In Crichton's imagined continuation of his documented journeys, this real traveler physically encountered humanlike "Wendols," or "mist monsters." As described to him by his companions, these creatures were hairy, foul-smelling cannibals who spoke no known form of language but who communicated readily with each other. And in the flesh they proved to be if anything even more appalling: their "shapes [were] hardly in the manner of men and yet also manlike," and in their presence "the air stank of blood and death" (143). Their hands were huge, well out of proportion to their arms, which like the rest of their bodies were covered with dark, matted hair. Their heads were enormous, the "faces... very large, with mouth and jaws large and prominent," and surmounted by "large brows... of bone" (245). This description is as graphic as any to be found in the by now enormous literature of nearly-but-not-quite-human creatures; and Crichton makes it clear in pseudo-scholarly footnotes and endnotes that the bizarre beings encountered by Ibn Fadlan in the fantasy portion of his journey are intended to be Neanderthals. Crichton's seriousness in his depiction of these creatures is emphasized by his dedication of *Eaters of the Dead* to the late Bill Howells, a revered Grand Old Man of paleoanthropology with whom he once studied, and who in his day was a noted Neanderthal scholar.

Clearly, through his evocation of the Neanderthals in *Eaters of the Dead*, Crichton was making a conscious play to all of the doubts and fears and conflicting feelings inherent in humankind's equivocal relationship to the world around it. Part of that world, and yet feeling that somehow we are separate from it, we are simultaneously attracted and repelled by anything that threatens to bridge the gap between ourselves and the rest of Nature. This makes for a powerful fictional device, yet one that can easily be overextended. Which of course raises the question of what one may reasonably make of Crichton's evidently deeply felt portrayal of the Neanderthals. Let's look briefly at these hominids, and at the highly biased history of their portrayal—by scientists, let alone by novelists—so that you may judge for yourself.

The Brute Emerges

The Neanderthals were the first extinct humans to enter the paleoanthropological pantheon, back in 1856. In August of that year, a crew of lime miners was hewing sediment out of a tiny cave in western Germany's Neander Valley ("Neander Thal"), overlooking the river Düssel near its confluence with the Rhine. In these sediments the miners found some fossil bones that they initially took to be those of a cave bear, an impressive but not unprecedented finding in those parts. Still, the foreman put them aside to be looked at by a local teacher and natural historian, Johann Fuhlrott. And, amazingly, Fuhlrott recognized these bones for what they were: the remains of a humanlike creature, the like of which had never been seen before. Not so, alas, most of the learned professors who eventually pronounced on this remarkable find. In 1856 the publication of Charles Darwin's great book, *On the Origin of Species by Natural Selection*, was still three years in the future; and in those pre-evolutionary times available scientific explanations for the remarkable phenomenon of the Neanderthaler were strictly limited. For although the few bones of the body skeleton recovered by the lime miners were recognizably human, the skullcap that went along with them was strangely unfamiliar. True, it had contained a large brain, fully as large as that of a modern person. But this brain had been enclosed within a skull of highly unusual shape. Unlike the high, rounded modern skull, with its forehead rising sharply above smooth brows, the skull of the Neanderthaler was long and low, and was adorned at the front with prominent brow ridges that arced above each eye socket. These latter features recalled to many the brow ridges of apes, as in a very approximate way they do. But this superficial resemblance to apes sat ill with the large brain, the most prized of all modern human features.

What, then, to make of this bizarre combination? Hermann Schaaffhausen, the learned anatomy professor at the University of Bonn to whom Fuhlrott submitted the fossils, recognized that they were very old—they were strongly fossilized, and had been found along with the remains of extinct animals. He was also impressed by the robustness of the bones, and concluded that their possessor had been enormously strong. He noted the apelike nature of the large brow ridges; but ultimately he placed more emphasis on the large brain, and concluded that the bones had belonged to a member of an ancient and barbaric human tribe. Thus, right at the beginning, was born the myth of the Neanderthals as primitive brutes. Not that everyone was in agreement with this particular point of view. Schaaffhausen's retired faculty colleague August Franz Mayer took him to task on virtually every aspect of his analysis and offered an alternative, namely that the remains were those of a pathological modern human, a sufferer from childhood rickets. Indeed, Mayer was prepared to go farther. Noting an apparent bowing of the legs, he suggested that the individual had been a horseman. And he hazarded that the bony ridges above the eyes had simply been formed by a brow constantly furrowed in pain from a damaged elbow. Putting it all together, Mayer painted the picture of a wounded Mongolian deserter from the Cossack army that had swept through the region to invade France in the wake of the Napoleonic debacle of 1814.

Mayer's scenario was roundly derided by the English anatomist Thomas Henry Huxley, who wondered how a wounded Cossack could possibly have "divested himself of his accoutrements," "crept into a cave to die," and managed to bury himself in "loam two feet thick" (Huxley 435). Nonetheless, championed by Rudolf Virchow, Germany's top pathologist of the day, the notion of the Neanderthaler as somehow a pathological modern human maintained wide currency until, by the end of the nineteenth century, several similar finds at widely scattered sites in Europe had made it impossible to avoid the conclusion that here was a real biological entity, by implication to be understood in its own terms rather than as a bizarre version of something else.

Actually, by the time that Mayer made his pronouncement in 1864, Darwin's book had already opened the way for the interpretation of the Neanderthaler in evolutionary terms, that is to say as an extinct precursor or relative of *Homo sapiens*. But in those early years few scientific minds were ready to take the plunge, although the contrarian Dublin anatomist William King formally named the species *Homo neanderthalensis* as early as 1864. Even the combative Huxley, who won renown as "Darwin's bulldog" for his energetic defense of the notion of evolution, sided with Schaaffhausen on this matter and concluded that the Neanderthaler was an ancient and savage member of the human species—and in doing so did much to entrench among English speakers the notion of Neanderthal bestiality. This is perhaps because the idea fed into an image that was already part of English folklore: my colleague Richard Milner has recently drawn attention to fifteenth-century depictions of the "wodewose," or "wild man of the woods," a savage, bearded, wild-haired, club-wielding figure believed to roam the already quite densely populated landscape of southern England. This image hardly required any modification into the prototypical cave man; and once it had been established that the Neanderthals were indeed an independent phenomenon to be reckoned with, it was in this bestial light—the brutish, club-wielding savage towing his woman around by the hair—that they were typically portrayed. In contrast, and tellingly, when the fossils of ancient *Homo sapiens*—approximate contemporaries of the Neanderthals—began to be discovered a decade after the Neanderthaler, they were often represented as Rousseauesque "noble savages," Adam- and Eve-like figures inhabiting a Stone Age Garden of Eden.

Over the first half of the twentieth century, Neanderthal fossils continued to be discovered in a wide swath of Europe and western Asia. In the public mind these extinct human relatives continued to embody the very essence of brutishness: an impression that was powerfully reinforced, shortly before World War I, by the Paris anatomist Marcellin Boule's interpretation of a Neanderthal skeleton from the French site of La Chapelle-aux-Saints. According to Boule, the remains were those of a thick-necked, bent-kneed, and shuffling evolutionary dead-end: the perfect picture of a benighted primitive.

Shortly after this, *Homo neanderthalensis* began to become lost within a thicket of new names that were proposed for practically every hominid fossil discovery that came along. But at around the time, in mid-century, when a general simplification of names placed *Homo neanderthalensis* back at center stage, it turned out that the Neanderthal individual so influentially described by Boule had been stricken with an advanced case of arthritis; and once scientists had begun to recognize this, the pendulum began to swing back with increasing momentum. Before long it had become received wisdom that, if he were appropriately showered, shaved and dressed, a Neanderthal would pass entirely unremarked on the New York City subway. Thus, in came these curious hominids from the cold: from the wilds of separate species status, the Neanderthals were welcomed back into the bosom of *Homo sapiens* as the subspecies *Homo sapiens neanderthalensis*. That sounds fine; after all, the Neanderthals had brains as large as ours, and in the liberal-minded post-war years it seemed only fair to admit these highly encephalized hominids to fully human status. Indeed, it seemed rather disgracefully discriminatory not to. The only problem is this: that, when you are absorbed, you tend to lose your identity. And that is, in fact, what happened to the Neanderthals, who never stood a chance of being properly understood as long as they were viewed as no more than an odd—and by implication inferior—variant of ourselves.

So Who Were the Neanderthals?

It's been known for years—indeed, ever since 1856—that the Neanderthals, however you classify them, were very different physically from us. Their big brain was housed in a skull of dramatically different shape from ours; and the individual bones of the body skeleton, with their thick-walled shafts and clunky joint surfaces, have always been readily distinguishable from our own. Yet no complete Neanderthal skeleton has ever been found, and it came as quite a revelation to see the first composite bony reconstruction of an entire Neanderthal. This was a veritable Frankenstein, recently cobbled together from bits of a half dozen partial skeletons by my American Museum of Natural History colleagues Gary Sawyer and Blaine Maley. And it revealed an individual strikingly different in its body proportions from us. The distinctions are most dramatically evident in the thoracic and abdominal regions. In contrast to our own rib cage, which is more or less barrel-shaped and tapers inward at the bottom as well as at the top, that of the Neanderthal flared dramatically outward from a narrow top to meet the wide pelvis at a very short waist. These differences reflect a being that would not only have *looked* decidedly distinctive on the landscape, but whose stiff-waisted gait would have been different from ours, too. The contrast is sufficiently striking to make it highly unlikely that members of the two groups would have found each other attractive as mating partners.

Interestingly, given the hulking image of the Neanderthals that stems all the way back to Hermann Schaaffhausen's original description, the new skeleton is not that of an enormously powerful bruiser. Certain elements, such as the finger bones, carry the traces of very powerful flexor muscles; but in many other respects this individual does not look particularly Herculean. This is not the muscle-bound goon of popular mythology.

The new reconstruction of the Neanderthal skeleton is, however, consistent with the notion that the Neanderthaler and his kind belonged to a species entirely distinct from ours. Virtually all the other biological evidence points this way, too, including recent indications from the Neanderthal genome. DNA, the long, twisting molecule of heredity, is notoriously fragile and rapidly breaks up into ever-tinier fragments after an organism dies. However, amazing recent advances in molecular genetics have allowed the recovery of bits of the Neanderthal genome. These show that, while Neanderthals varied among themselves in their DNA—just as we do—they varied around a distinctly different mean. What is more, molecular geneticists have not been able to find any convincing traces of the Neanderthal genotype in any living human population. Indeed, the molecular differences between the two species are such as to suggest that modern humans and Neanderthals last shared an ancestor well over half a million years ago—and this finding is comfortably consistent with what a rapidly expanding fossil record seems also to be telling us. Recognizably Neanderthal fossils only go back to around 200,000 years ago, which is plausibly around the time the species *Homo neanderthalensis* was born; but related fossils that were clearly not our own ancestors are known back almost to the half-million year mark, and suggest that the larger groups to which *Homo sapiens* and *Homo neanderthalensis* belong have been separate for at least that long. The two hominids are/were members of entirely different lineages, not simple variants of one another.

Well, what were the Neanderthals like as living beings? Is there any evidence that they shared any of the cognitive uniquenesses that distinguish us from all other organisms in the world today? Sadly, only a tiny part of the Neanderthals' behavior is reflected in the stone tools and detritus of living that the Neanderthals left behind. But overall it is hard to avoid the impression that they perceived and reacted to the world around them in a way that differed significantly from our own. The contrast comes across most clearly in the comparison between the behavioral record of the Neanderthals and that of the Cro-Magnons, the *Homo sapiens* who invaded their European heartland in the millennia following about 40,000 years ago—and who entirely evicted them in the space of 10,000 years. Nobody knows exactly how this eviction—this extinction—was achieved. Maybe it was through direct conflict; maybe it was the result of indirect economic competition. Perhaps it was even just coincidence; it has been argued that the Neanderthals, already thin on the ground, simply succumbed to the rigors of the approaching Ice Age. What we can be pretty sure of, though, is that the Neanderthals were not simply genetically absorbed by hordes of incoming Cro-Magnons. Of course, we're not in a position to deny absolutely that some kind of Pleistocene hanky-panky might not have occurred, especially given the bizarre proclivities of some *Homo sapiens* today; but in light of the virtual (though still ritually debated) certainty that the Neanderthals and Cro-Magnons belonged to different species, the chances that any biologically significant intermixing took place are slim to none. There is a small amount of archaeological evidence, limited in time and space, that some Neanderthals may, regionally, have by some means acquired a few of the new technological and other behaviors that marked the Cro-Magnons; but in most places the Neanderthals appear to have been suddenly replaced by Cro-Magnons, their way of life abruptly terminated.

One conclusion that you might draw from this is that the Neanderthals were in some way inferior to the Cro-Magnons. And in the limited sense that—for whatever reason—they appear to have been ecologically uncompetitive with the newcomers, this appears inarguable. But it would be totally unjust to expand this notion to conclude that they were inferior versions *of ourselves*. They were simply *different*. And this makes it impossible for us to put ourselves in their place, for we human beings are totally incapable of imagining any cognitive state other than our own. Observing a chimpanzee sitting behind the bars of a zoo, we may surmise that it is rather dejected—as indeed it may well be. But what does this really mean about its subjective experience? It is easy to tell ourselves that the chimpanzee is unhappy—as it very probably is. But what it is thinking—or whether it is thinking at all in any sense we would recognize—we simply can't know. Still less can we read from the paleontological and archaeological records how the Neanderthals—so much closer to us than chimpanzees are, yet obviously not the same—experienced the world. Chances are, though, that they, too, saw the world through alien eyes—that is to say, through eyes more similar to other denizens of nature than to our own.

Why do I say this? The Neanderthals were skilled makers of stone tools, and were almost certainly sophisticated hunters and exploiters of the environment around them. Indeed, recent research suggests that they may have specialized in the hunting of fearsome large-bodied mammals such as woolly mammoths and rhinoceroses. They were, indeed, a force to be reckoned with. But in a pretty large archaeological record they left few if any traces of symbolic thought or behaviors. And it is symbolic reasoning that separates us human beings most fundamentally from all of the other creatures that surround us. Modern humans alone, it appears, decompose the world around them into mental symbols that they can shuffle and reorder in their brains to make new mental worlds and ask questions such as "What if?" The Neanderthals, though they invented the practice of burying the dead, and clearly looked after the disabled, showed no unequivocal signs of symbolism. On the other hand, the Cro-Magnons who so rapidly ousted them from their European homeland unquestionably possessed the entire panoply of symbolic thought. Well over 30,000 years ago they were painting powerful art on the walls of caves—art to rival anything that has been produced since. They played music on vulture-bone flutes, and made notations on bone plaques. They made some of the most delicate and elegant carvings and engravings ever made. They baked clay figurines in kilns, and sewed clothing with tiny-eyed bone and antler needles. Their lives were drenched in symbol, and they were clearly *us*.

The acquisition of symbolic reasoning actually appears to have occurred well *after* anatomically modern humans first showed up in the fossil record. The origin of our species as a recognizable anatomical entity seems to have occurred close to 200,000 years ago. Yet as far as we can tell from current evidence, modern behavior patterns do not go back much if at all beyond about 75,000 years ago. What this strongly implies is that, while the neural hardware that permits symbolic thought most plausibly dates back to the origin of *Homo sapiens* as a distinctive physical entity, this capacity had to be discovered by some behavioral means. Quite possibly this was the invention of articulate language—the most quintessential of symbolic activities. And there is no convincing evidence that Neanderthals ever acquired a developed symbolic propensity, whether or not they possessed the potential to do so.

None of this is to make any value judgment about the Neanderthals. They were clearly remarkable hominids, resourceful and flexible in the face of environmental change. They doubtless possessed a sophisticated communicative ability, even if not language. They were highly social, and were skilled stone toolmakers. They controlled fire, built shelters, and almost certainly possessed material cultures that went far beyond what we can see in the preserved lithic record. They survived for an extended period in environments that were sometimes harsh indeed, and were probably able to exploit the environment around them more effectively than any of their predecessors had been. Clearly, they were cognitively complex. Indeed, in all likelihood they would be living today had they not encountered the entirely unanticipated phenomenon of *Homo sapiens*. But when that encounter occurred, two entirely different kinds of hominids met. One was the most sophisticated extrapolation yet of pre-existing trends in hominid evolution. The other—us—was something entirely new and unanticipated: a product, it seems, of a chance combination of old and new that gave rise to a creature of unprecedented cognitive function.

Wendols, Neanderthals, and Us

Had the words of *Eaters of the Dead* all been Ibn Fadlan's, they would be astonishing and revelatory. So much Wendol detail is there: big heads, large hands, strong bony ridges above the eyes, protruding faces. Yes, on the face of it these could truly *be* Neanderthals. But these are all features that we can read directly from the bones that the Neanderthals have left us. All of the other Wendol details, the most intimately personal and compelling ones—the wild, ragged hair, the pungent odors, the nocturnal and aggressive habits, the inchoate ramblings—are, alas, products of Crichton's imagining, not of Ibn Fadlan's observation. And it is precisely that kind of detail that we most want to know about the Neanderthals. We know at least in broad strokes what Neanderthals *looked* like; but what would it have been like to *meet* one? To be breathed on by one, looked in the eye by one, vocalized at by one? These intangibles are the things we most desire to know about these close relatives, so near to us and yet so far. In taking his portrait of the Neanderthals far beyond the facts, Crichton clearly aimed to help. And the feeling with which he writes carries the signal of conviction. But since we, too, know what Crichton knew when he wrote this, the edge is somehow off his account. A fictional Ibn Fadlan, the mental product of a writer belonging to *Homo sapiens*, simply cannot help us to know any more about the Neanderthals than the basics we can learn from science. This is, of course, because the novelist's approach inevitably leaves us with an artifact of the human mind, rather than with a genuine invention of Nature. It tells us a lot more about us than it does about the Neanderthals. What we see in the Wendols and in visions like them is a dark and gloomy image of ourselves, a product not of the external world but of our own murky and atavistic doubts and fears. Still, by evoking unsettling responses such as these it may well be that the long-vanished Neanderthals, whether fictionalized as Wendols or not, continue to this very day to provide us with the truest available mirror in which to perceive not only our own uniqueness, but our ever-present uncertainties about who we truly are.

IAN TATTERSALL is a curator in the Division of Anthropology at the American Museum of Natural History in New York City. His research is in paleoanthropology and lemur biology, and he is the author of some twenty books including *Becoming Human* (Harcourt Brace, 1998), *Extinct Humans* (with Jeffrey Schwartz: Westview Press, 2000), *The Monkey in the Mirror* (Harcourt, 2002), and the forthcoming *New Oxford World History, Vol 1: Beginnings to 4000 BCE* (Oxford University Press) and *Human Origins: What Bones and Genomes Tell Us About Ourselves* (with Rob DeSalle: Texas A&M University Press). He lives in Greenwich Village.

References

Crichton, Michael. *Eaters of the Dead* [enlarged version]. New York: Avon Books, 1992.

Huxley, Thomas Henry. "Further Remarks on the Human Remains from the Neanderthal." *Natural History Review* (London) 4 (1864): 429–446.

PRIMATE BEHAVIOR AND MISBEHAVIOR IN MICHAEL CRICHTON'S *CONGO*

Dario Maestripieri, Ph.D.

When Congo *was published, according to Michael Crichton, most of the book's reviewers found Amy—the sign-language-using gorilla—to be unbelievable, despite being based upon the real-world gorilla Koko. Since Koko's rise to fame, other intelligent animals have gained notoriety. The language skills of African Gray parrots N'kisi and the late Alex are legendary. A chimpanzee named Washoe, like Koko, was trained in American Sign Language. Despite impressive accomplishments in the field of animal learning and behavior, is it reasonable to expect that a human-trained gorilla could be our interface with wild gorillas? That's what we asked Dario Maestripieri, Ph.D.*

AS EVERY BUSINESS EXECUTIVE KNOWS only too well, good ideas can generate a lot of money. Bad ideas sometimes generate money too, but usually at the expense of someone else. Here is an idea: let's teach chimpanzees and gorillas human language and take them back to the jungle so that they can serve as interpreters for other apes and help us understand Mother Nature's secrets. This seemed like a good idea when it first crossed people's minds about 100 years ago, but turned out to be a bad one. Nevertheless, this idea generated a lot of money for some folks: the researchers who obtained millions of dollars from the U.S. government to attempt to teach language to apes, and Michael Crichton, who used it to write *Congo* and made huge profits from the sales of his book and its movie rights. In the end, this business venture was arguably more legitimate for Michael Crichton than for the ape language researchers. *Congo* is an excellent book in its genre, its success was fully justified, and the folks who spent a few bucks to read it probably thought that their investment was well worth it (the movie *Congo* sucked, but that's another story). Chimpanzees and gorillas, however, have never learned to speak English, we've never had any intellectual conversations with wild apes, and even if one day we figured out a way to talk with them, I'm not sure we would learn anything interesting about Mother Nature that we don't already know. If anything, we could probably teach the apes a few things about themselves that they don't know—assuming that they would care to listen, which I strongly doubt. Although we have learned some interesting things from our unsuccessful attempts to teach spoken English to the apes here in the U.S., many folks these days believe that the millions of dollars of taxpayers' money invested in this research could have been used to build new schools instead.

In writing *Congo*, Michael Crichton did his homework well and researched the history of ape language studies. His telling of the story is pretty accurate, although a few bits and pieces are missing here and there. Somewhere in one of *Congo*'s early chapters, Crichton mentions Keith and Kathy Hayes, the Florida couple who in the early 1950s raised a chimpanzee infant named Vicky at home with them. He doesn't mention that another husband and wife research team—the Kelloggs—had tried the same thing a few years earlier, and raised an infant chimp named Gua at home along with their own son Donald. In some psychology textbooks, one can find hilarious photos taken by the Kelloggs, which show Gua and Donald holding hands and wearing the same pajamas before going to bed. Both the Hayeses and the Kelloggs tried to teach spoken English to their infant chimpanzees. To understand why they thought this was a good idea, one needs to know that ape language research began during the heyday of behaviorism, a particular brand of psychology that maintains that behavior—human or animal—is the product of the environment and that almost anything can be taught to anybody. Raise an ape in a home in the suburbs and it will turn into a person. Speak to them in English and, at some point, they will speak back to you, just like every child does. Well, neither Gua nor Vicky ever spoke back to their foster parents. Chimpanzees and other apes simply don't have the right vocal cords to produce sounds like human speech, and even if they did, they could never learn to speak English fluently, or any other human language. Many aspects of human behavior—including language, according to the linguist Noam Chomsky and his followers—are not learned but genetically inherited. Other behaviors are learned from the environment, but we probably have strong genetic predispositions to learn them the way we do. The same goes for the behavior of other animals, of course. Raise a young chimpanzee in a home in the suburbs and the result will be a screwed-up chimpanzee, not a human. And if you have seen the film *American Beauty*, you know that human children raised in those homes may turn out to be pretty screwed-up too.

After the failures to teach spoken English to Gua and Vicky, other folks used different strategies, for example, teaching apes sign language, or communicating with them using symbols on a keyboard. Crichton mentions the names of some of these famous apes: the chimpanzees Washoe, Lana, Sarah, and Nim Chimpsky (an obvious joke on Noam Chomsky), and Koko the gorilla. He forgot Chantek the orangutan, but made up for it by inventing the names and stories of many other language-trained apes. Too bad *Congo* was written before the most extraordinary talking ape of all times—Kanzi the bonobo—entered this unusual hall of fame. In fact, at the time he wrote *Congo*, Crichton didn't seem to know that bonobos or pygmy chimpanzees—another African ape species similar to chimpanzees—even existed. Although bonobos were discovered in 1928, in *Congo* we read that there are only two species of African apes: chimpanzees and gorillas.

One thing Crichton seemed to know that I didn't is that the U.S. Air Force began funding ape language research in the 1960s as part of a secret project called *Contour*. This project—according to Crichton—involved the development of strategies to be used in case of contact with alien life forms from outer space. Teaching apes language and taking them back into the field to serve as interpreters was viewed by U.S. Air Force strategists as a stepping stone toward the future use of these primates as intergalactic ambassadors. Why take any chances trying to establish conversations with potentially hostile aliens equipped with head-chopping light sabers when chimpanzees and gorillas could be sent out there on the starship *Enterprise* to negotiate with the aliens on our behalf? Sounds reasonable to me, although animal rights activists would probably object to this use of our ape cousins. Since it's true that NASA and the U.S. Air Force gave some researchers a lot of money to teach monkeys and apes computer video games that simulated piloting war jets and launching missiles (another project that wouldn't get the stamp of approval of animal rights activists, I'm afraid), I wouldn't be surprised if NASA also funded projects preparing close encounters of the third kind between apes and aliens. As for this specific *Contour* project mentioned by Crichton, I've never heard of it. An Internet search using the Google engine revealed that NASA indeed had a project called *Contour*, but this project involved the use of a remotely guided spacecraft—and not gorillas—for the exploration of comets. Moreover, this project was launched in 2002—long after *Congo*'s first edition (1980). But maybe this project had a secret component we don't know about, which was inspired by Crichton and his book. When he wrote *Congo*, Crichton was certainly optimistic about the future outcomes of ape language research. He predicts that in the near future signing apes will be called to testify in court during custody cases involving themselves or their fellow primates. That hasn't happened yet and probably never will. Crichton was more accurate in predicting the invention of the Internet: in *Congo*, he predicts that by the year 1990 there will be one billion computers linked by communication networks to other computers around the globe.

In any case, *Congo* tells the story of a gorilla named Amy, who has been taught American Sign Language and is taken back to her native jungle in Central Africa in hopes that she will communicate with the local apes. To make the story even more interesting than the *Contour*-funded projects, the local apes Amy is supposed to talk with are not peaceful leaf-eating mountain gorillas but a new species of gorilla, who like to kill people by crushing their skulls with hard rocks. And the purpose of the folks who accompany Amy on this trip to Africa is not to discover Mother Nature's secrets, but to find the lost city of Zinj and its precious diamonds.

As in real life, the ape language project fails. When Amy finally meets the murderous gorillas, they don't care to learn American Sign Language from her and she can't speak their invented language, made of wheezing sounds and hand gestures "delivered with outstretched arms in a graceful way, rather like Thai dancers." She comes to understand enough of the gorilla language, however, to allow her trainer and mentor—fearless primatologist Peter Elliot—to put together and broadcast over the airwaves a message for the gorillas. So, when the apes are about to kill everybody, Elliot in particular, all of a sudden they hear loudspeakers emitting wheezing sounds meaning "Go away. No come. Bad here." That's enough for them to stop, turn around, and walk back home. Good for Elliot and his buddies that the gorillas are bottom-line kind of guys, who don't care much about syntax. Amy saves Elliot's life one more time by hugging him and pretending that he's her child while he is under attack by the gorilla killing-machines. Apparently, these gorillas are smart enough to figure out how to deactivate a computer-controlled, laser-equipped defense perimeter put up around our heroes' camp, but so dumb that they buy into Amy's pathetic little act.

So, if the future of humankind had to depend on Amy's negotiations with hostile aliens who were about to blast our planet from the universe with their laser cannons, I wouldn't be too optimistic. She couldn't even communicate with dumb gorillas who smash people's heads with rocks. Who needs ape extra-galactic ambassadors, however, if what Crichton writes in *Congo* is true—that a guy named Seamans at the University of California at Berkeley had developed a computer program named APE—it stands for Animal Pattern Explanation—capable of "observing" Amy and assigning meaning to her signs? Since, Crichton argues, the APE program utilized declassified army software subroutines for code-breaking and was capable of identifying and translating new signs, "there was no reason why it would not work with an entirely new language," whether this be animal, human, or alien. No need for extragalactic ambassadors, then. When the mean aliens come and start waving their arms ominously at us, let's have the APE program figure out exactly what they are saying and help us deliver our response to their threats of global annihilation: "Go away. No come. Bad here."

Unfortunately, this Seamans guy and his magical language software existed only in Crichton's imagination. In *Congo*, Crichton mentions many real scientists' names and describes their research pretty accurately, but he mixes them up with invented characters all the time, and it's not easy to tell who's real and who isn't. To complicate things, he sometimes uses the name of a real scientist, for example, Rumbaugh—a well-known ape language researcher—for a character he invented. Crichton especially likes to report quotes from scientists making extraordinary claims. Take this one for example: "In 1975, the mathematician S. L. Berensky reviewed the literature on primate language and reached a startling conclusion. 'There is no doubt,' he announced, 'that primates are far superior in intelligence to man.'" A Google search for "S. L. Berensky" produced some interesting results. The first hit is a Russian Web site where you can read the whole *Congo* book online—so much for copyright laws. Most of the other hits are porn Internet sites in which sentences lifted from well-known books are mixed with mumbo-jumbo text, commercial ads, and links to other porn sites. Apparently, the word combination "S. L. Berensky reviewed" is very popular among e-mail spammers: if you check your own mailbox, you might find messages that have it as their subject title, and offer you a penile enlargement procedure at a bargain price. Needless to say, there is no trace of the real mathematician S. L. Berensky on the Internet, or of his primate language studies and their startling conclusions. This other quote found in *Congo* almost fooled me until I noticed a pattern in the names of the experts Crichton likes to cite: "L. S. Verinsky once said that if alien visitors watched Italians speaking, they would conclude that Italian was basically a gestural sign language, with sounds added for emphasis only." Again, a Google Internet search did not produce any traces of L. S. Verinsky or his linguistic wisdom.

Clearly, while researching his material for *Congo*, Crichton must have talked with a lot of researchers and maybe even attended a couple of scientific meetings. Otherwise, how would he know that biologists who study animal behavior like to wear jeans and plaid lumberjack shirts at their conferences? And I am not sure whom he talked with, but he certainly came away with a pretty cynical view of scientists and their world, especially when it comes to funding for their research. In *Congo*, Peter Elliot is described as a skilled grantsman, "someone who had long ago grown comfortable with situations where other people's money and his own motivations did not exactly coincide.... A researcher promised anything to get his money." Crichton also seems to have issues with cocktail parties attended by scientists in Houston. When Amy the gorilla first meets Karen Ross, the beautiful but cold-hearted computer expert on the team, "she went directly to her, sniffed her crotch, and examined her minutely." Ross minimizes the incident to Elliot by saying to him, "It was just like any cocktail party in Houston. I was being checked out by another woman." Somehow, crotch-sniffing doesn't seem to happen at the cocktail parties for scientists I've been to.

As for the behavior of gorillas and other nonhuman primates, Crichton knows a lot about that too. Here are some examples of the primate behavior facts you find in *Congo*: Gorillas like to make their own nest every night and sleep in them. When mountain gorilla silverback males attack an intruder, they go through a typical behavioral sequence including grunting, sideways movement, slapping, tearing up grass, beating chest, and charging. Male baboons often end their fights when one male grabs an infant and clutches it to his chest; the presence of the infant inhibits further aggression from the other male. Monkeys and apes spend a lot of time grooming each other to reduce tension. Chimpanzees in cages like to throw their feces at people, whereas gorillas like to eat them. Wild chimpanzees use twigs for fishing insects out of their nests, and youngsters learn fishing skills from their mothers in the course of observations and practice sessions that last several years. Chimpanzees are more aggressive and more into dominance hierarchies than gorillas, and more dangerous to people. Crichton also mentions that chimpanzees pose a threat to human children, so that when Jane Goodall studied chimpanzees at Gombe National Park in Tanzania, she had to lock away her own infant to prevent his being taken and killed by the chimps. I've heard these stories too. The statement that chimpanzees occasionally kidnap and eat human infants, however, seems a little exaggerated.

The accuracy of Crichton's understanding and description of nonhuman primates and their behavior begins to break down when he talks about their cognitive skills. For example, in the context of describing chimpanzees' termite-fishing skills, Elliot/Crichton confidently states that conspecific teaching is quite common among primates. He cites as an example the female chimpanzee Washoe who supposedly taught American Sign Language to her infant son Loulis. Yes, I've read that too, but because no real data were ever presented about Washoe's presumed lessons to Loulis—just somebody's anecdotal observations—I don't believe it. Crichton goes on to claim that language-skilled primates freely teach other animals in captivity: "They also taught people, signing slowly and repeatedly until the stupid uneducated human person got the point." Another jab at scientists, perhaps? Judging from the way Amy behaves in *Congo*, one would have to assume that gorillas and other apes have the ability to recognize themselves in mirrors and to imitate other individuals' actions. For example, at some point in the book, Amy uses a mirror to put lipstick on, and she buckles her seat belt in a car after observing how people do it. In reality, whether primates have a true sense of self and can imitate others' actions is still being debated in scientific circles. If anything, studies have shown that chimpanzees can learn how to use a mirror to wipe something off their forehead or clean their teeth, but gorillas seem to have a lot more trouble doing it. Although this in no way implies that gorillas are dumber than chimpanzees, Crichton's statement that "Now there was abundant evidence from field and laboratory studies that the gorilla was in many ways brighter than the chimpanzee" is certainly not warranted either.

Crichton definitely goes overboard on the issues of primate dreams and their understanding of time. In *Congo*, we learn that Amy was the first primate to report dreams. We are told that a manuscript written by Elliot and titled "Dream Behavior in a Mountain Gorilla" was submitted for publication and reviewed by three experts but never accepted for publication. We are not told what the reason for the rejection was, but it may have had something to do with Elliot's explanation that gorilla dreams represent a form of genetic memory, which, in Amy's case, brought back to her mind images of the jungle in which she and her fellow gorillas were born and brought up, and that had become somehow encoded in her DNA. Elliot finds help from Freud in explaining Amy's dreams. He concludes that the dream protected Amy from a situation that had to be changed, but that Amy felt powerless to alter, especially considering whatever infantile memories remained from the traumatic death of her mother. Just like behaviorism, Freud is not so popular among psychologists these days.

In *Congo*, we also learn that Amy distinguishes past, present, and future because she remembers previous events and anticipates future promises. So far so good: every animal on the planet with a brain larger than a single neuron can do that. Here comes the tricky part, though. According to Elliot/Crichton, Amy's behavior seemed to indicate that she conceived of the past as in front of her—because she could see it—and the future behind her—because it was still invisible. Therefore, whenever she was impatient for the promised arrival of a friend, she repeatedly looked over her shoulder, even if she was facing the door. Now, THAT is an original, bold, and imaginative suggestion. If that were true, it would explain why so many mountain gorillas die when they get shot at by poachers in Central Africa. When they see a human in front of them pointing a gun and taking aim at them, instead of looking ahead and trying to dodge the bullet, they turn around and look over their shoulder. No wonder they are on the brink of extinction!

Aside from this gem, the discussion of Amy's intelligence in *Congo* is generally less interesting than the description of her misbehavior. Clearly, Crichton seems to like bad girls' behavior and makes sure Amy has it all. In addition to drinking martinis and champagne, smoking cigarettes (but only to relax), and sniffing women's crotches, Amy swears too. The way she does it is she first signs the name of the person she wants to insult—Peter, for example—and then taps the undersides of her chin—the sign commonly used to communicate the need to go to the potty. Crichton/Elliot remarks that primate investigators were under no illusions about what the animals really meant in these circumstances: Amy was saying *Peter is shitty*. Crichton goes on to confidently inform his readers that nearly all language-trained apes swore. According to him, "at least eight primates in different laboratories had independently settled on the clenched-fist to signify extreme displeasure, and the only reason this remarkable coincidence hadn't been written up was that no investigator was willing to try and explain it. It seemed to prove that apes, like people, found bodily excretions suitable terms to express denigration and anger."

Well, Michael Crichton, you certainly got one thing right: there haven't been any scholarly publications reporting on the swearing behavior of language-trained apes. As for the story about the evolution of the ape clenched-fist gesture, I find that hard to believe. If you had claimed, however, that after having been put through the ordeal of trying to learn language for all these years, apes in different laboratories had independently learned how to give their trainers the finger, I might have believed you.

DARIO MAESTRIPIERI earned his Ph.D. in psychobiology from the University of Rome, Italy, in 1992. He is currently an associate professor of comparative human development and evolutionary biology at the University of Chicago. His research interests focus on the biology of behavior, and in particular on physiological, ecological, and evolutionary aspects of primate social behavior. Dr. Maestripieri has published over 130 scientific articles and several books including *Primate Psychology* (Harvard University Press, 2003) and *Macachiavellian Intelligence: How Rhesus Macaques and Humans Have Conquered the World* (The University of Chicago Press, 2007).

WE STILL CAN'T CLONE DINOSAURS

Sandy Becker

Michael Crichton's version of Frankenstein *is one of his most popular novels, and it spawned the only sequel he has ever written as well as three movies (more are being planned). Is the cloning of dinosaurs from ancient DNA even possible, though? Sandy Becker examines the technical hurdles in creating a real-world* Jurassic Park.

DOLLY, THE CLONED FINN DORSET LAMB born in Scotland in 1997, is surely the most famous sheep on the planet since she was the first clone to be followed by the popular press. Frogs and mice had been cloned years before, and those feats were of great interest to the scientific community, but invisible to the general public. Dolly (who died in 2003) was a large, fluffy, familiar animal. She was cloned by Ian Wilmut and his colleagues at an agricultural research facility in Scotland, using a process known as "nuclear transplantation"—the entire nucleus of the donor cell, the packaged DNA, and all its paraphernalia were transplanted into a sheep's egg whose own nucleus had been removed. Thus, Dolly is the genetic replica, or clone, of the animal who provided the nucleus, not the animal who provided the egg. The donated nucleus came from a mammary cell, which prompted the lamb's creators to name her after the country western singer with the hitherto most famous mammary glands.

Long before there was a real clone, however, there were dozens of fictional clones cranked out in dozens of novels: some eloquent, some exciting, some scientifically plausible, some none of the above. Perhaps the most famous are the dinosaurs of Michael Crichton's 1990 novel *Jurassic Park*. Resurrected from scraps of dinosaur DNA rescued from the stomachs of mosquitoes that had been trapped and preserved in amber just after feasting on dinosaur blood some 100 million years ago, these unpredictable creatures clomped their way through an engrossing book and a high-grossing movie—and their sequels.

The story in *Jurassic Park* is propelled by the concerns of investors that the park, featuring live cloned dinosaurs, must be safe. The owner, millionaire John Hammond, brings a team of inspectors to the park, including two paleontologists, Alan Grant and his assistant Ellie Sattler, and Ian Malcolm, a mathematician specializing in Chaos Theory. There would be no story if nothing went wrong, so of course a number of things do—the cloned dinosaurs are unexpectedly ferocious, the computer systems in the park fail, the dinosaurs apparently escape from the island on which the park is located. Quite a few people die. These problems provide a series of hair-raising adventures for the characters, and a number of opportunities for the author to explore some scientific issues that interest him.

In the real world, cloning—that is, nuclear transplantation—has been done for decades. Dolly is the first famous clone, but she is by no means the first multi-cellular clone. John Gurdon cloned several frogs in 1966 by transplanting nuclei from the cells lining a tadpole's intestine into eggs that had been irradiated to destroy their own nuclei. Since then mice, pigs, cattle, cats, horses, goats, a dog, and of course Dolly have been successfully cloned. The list of cloned animals on Wikipedia now includes sixteen entries. It has turned out to be quite a bit more challenging to clone mammals than to clone frogs. For openers, frog eggs are large, numerous, and sturdy, and embryonic development normally takes place in ordinary pond water. The donor nucleus can be literally "injected" into the recipient egg with a micro-injector. Mammalian eggs are tiny, few, and fragile, and the cloned embryo must be returned to a surrogate uterus to gestate. The recipient egg must usually be tricked to open and accept the donor nucleus by a virus or a jolt of electricity. Since people are mammals, they are generally more interested in mammals, and therefore cloning research has primarily focused on them.

The pioneer cloners in the real world have had to solve a number of problems. How do you get the donor nucleus to act like the nucleus of a newly fertilized egg, rather than the nucleus of the body cell it used to inhabit? During the growth of an individual from fertilization to adulthood, the DNA in each cell undergoes changes that enable it to carry out the business of running that cell. Most of the changes are permanent. Most of the cells in the body of an adult do not divide, for example. A few specialized cell types, called "stem cells," provide a continuous stream of replacements for worn-out cells, but most cells have the genes that control cell division permanently switched off. How can such DNA be "reprogrammed" to act like embryonic DNA? The first thing a normal egg does when fertilized is begin to divide into two, four, eight, and soon hundreds of cells. A nucleus transplanted into a waiting egg must somehow be synchronized with the division cycle of the egg, lest for example the DNA be apportioned into two cells before it is finished replicating a second copy, or, equally disastrous, the DNA replicates more than one copy of itself before cell division takes place. Is it better to use an unfertilized egg, an oocyte, to receive the transplanted nucleus, or an already fertilized zygote? How best to perform the delicate task of removing the nucleus of the recipient cell and inserting the donor nucleus? Does it matter at what stage in the cell division cycle the recipient cell and donor nucleus are when the transplantation is done? The work of Ian Wilmut and many colleagues and predecessors has gone a long way to answering these questions, but they are still the focus of ongoing research.

Cloning by nuclear transplantation is still pitifully inefficient, with researchers chalking up hundreds of failures for each successful clone. It took Ian Wilmut 277 tries to clone Dolly. The details of his failures are even more daunting. The 277 represent "fused couplets," enucleated eggs which could be seen in a low-power microscope to have fused with a donor cell carrying the DNA. This step is apparently pretty successful, as 64 percent of his egg/donor pairs fused. However, only 12 percent survived development long enough to become a blastocyst (hollow ball containing dozens of cells), ready to implant in the surrogate ewe. The twenty-nine embryos that successfully developed into blastocysts were implanted in thirteen ewes, only one of which (Dolly's mom) became pregnant.

Undaunted, in 2000 Lou Hawthorne, Mark Westhusin, and venture capitalist John Sperling founded a company called Genetic Savings and Clone, which offered to clone your beloved pet for $50,000. The company, based in California, actually did clone a few cats for paying customers, but has since gone out of business. Even at $50,000 a pop, they weren't making a profit.

Despite the inefficiency of the nuclear transplantation process, in some ways cloning extinct animals doesn't seem so farfetched. In 1994, a few years after *Jurassic Park* was published, Scott Woodward and a team of scientists in Utah demonstrated that it might just be possible to get DNA out of dinosaur bones. Recently (April, 2007), a multidisciplinary team extracted some identifiable collagen (the main protein component of bone and cartilage) from a specimen of Tyrannosaurus Rex found in Montana. Their article was published in *Science*, a prestigious, peer-reviewed scientific journal that does not generally publish the work of crackpots. In 1996 Hendrik Poinar and an international team of researchers published an article in *Science* showing that amber is indeed an excellent preservative for ancient DNA (although probably not good enough to yield well-preserved dinosaur DNA).

DNA has been successfully extracted from fossils somewhat younger than those of dinosaurs. In 2007 a French team compared museum fossils of various species with freshly excavated specimens up to 50,000 years old, and found that far more DNA could be recovered from freshly excavated specimens than from those stored in museums.

In 1997 a team of German scientists extracted DNA from a Neanderthal specimen, and since then over a dozen articles have been published investigating Neanderthal DNA (mostly showing that modern humans did not interbreed with Neanderthals before they died out). By 2006 a million base pairs of Neanderthal DNA had been sequenced. This may sound like a lot, but the human genome has 3 billion base pairs, so they are only 1/3000th done. So far, the Neanderthal genome is 95.5 percent identical to ours, a little more like us than chimps, who are only 95 percent identical to us.

And there are ongoing attempts to clone extinct or nearly extinct animals. A Galapagos tortoise nicknamed Lonesome George, who may be the last of his subspecies, has been suggested as a candidate, and samples of his tissue have been safely frozen for cloning. In 2000, workers at the biotech company Advanced Cell Technology cloned a gaur, an endangered Asian ox, by transferring skin cell nuclei from a recently deceased one into cow eggs. The experience underlines the difficulty of cloning mammals: ACT started with 692 cow eggs with inserted gaur nuclei; only eighty-one developed far enough to implant into the surrogate mom, and only eight cows became pregnant. (At this rate of success, the 238 cloned dinosaurs in Jurassic Park would have required starting material of 164,696 eggs with transplanted nuclei.) The one gaur embryo that survived to birth died two days later of common dysentery. Two years later, ACT used a similar technique to successfully clone a banteng, another endangered Asian ox. Researchers at the University of Teramo in Italy have cloned a mouflon, an endangered wild sheep. In each case the successful cloners have had access to well-preserved frozen tissue from which to collect the donor nuclei, and a large number of eggs available from a closely related domestic species, which also provided gestational surrogates.

Those hoping to clone already extinct animals are not so fortunate. They will be lucky to get viable DNA, let alone an intact nucleus, from their frozen or fossilized donor. It has been suggested by those prepared to give up on the possibility of cloning dinosaurs that perhaps mammoths would be a better bet. Several have been recovered from the permafrost and at least one has been chopped out and airlifted to a lab, ice and all. Elephant oocytes could serve as recipients for the donated mammoth DNA; certainly a better match than crocodile eggs for dinosaur nuclei! Surely an intact nucleus could be recovered? Probably not. Tissue that is promptly frozen in liquid nitrogen (about minus 200 degrees C) when it is barely dead can be preserved indefinitely, but the permafrost isn't THAT cold, and the natural process of dying and freezing slowly is unlikely to have preserved any intact nuclei for 20,000 years.

The quagga, a zebra-like animal whose last surviving member died in a Dutch zoo in 1883, has been suggested as a cloning candidate, since sloppy nineteenth-century taxidermy seems to have left some DNA-containing soft tissue still attached to stuffed specimens. And a Tasmanian Tiger pup preserved in alcohol may yield good enough DNA to enable Australian scientists to reconstruct viable chromosomes. However, these projects are stymied by the same problem that faced the dinosaur cloners in *Jurassic Park*: to clone something you need the whole nucleus, not just naked DNA.

The DNA of creatures larger than bacteria comes tidily packaged in a nucleus, complete with proteins to wrap and unwrap it and to help carry out its business. The DNA and proteins are not only encased in a "nuclear membrane," they are also arranged on a "nuclear scaffold" which keeps them from floating around willy-nilly inside the membrane. Each human cell contains DNA that would stretch to six feet long if it weren't wound up tightly. (Dinosaurs, of course, might have had more or less DNA). In animals from flies to lizards to humans, the DNA is organized into chromosomes, in pairs containing one from each parent. For example, humans have twenty-three pairs of chromosomes; mice have twenty, dogs have thirty-nine, fruit flies have four. Each chromosome contains hundreds to thousands of genes, and each nucleus contains not only the chromosomes but also proteins to package the DNA and to instruct each gene to get busy—or to do nothing. "Busy" for a gene means making messenger molecules (RNA) to send out of the nucleus, where they will be used as templates for making the many proteins each cell needs. Naked DNA, without this packaging and these proteins, can't carry out the business of running and replicating a cell, let alone the complex choreography of embryonic development. So, to clone a sheep or a cat or a dinosaur, you need the complete nucleus.

Michael Crichton, who holds an M.D.—although he veered off into fiction writing early on in his career—presumably knows of these problems. At the end of *Jurassic Park*, he acknowledges the members of the Extinct DNA Study Group at UC Berkeley, one of whom (George Poinar) was an author on the study described above on DNA preserved in amber. But Crichton's dinosaur cloners got only patched-together scraps of DNA, not any complete nuclei. The novel goes into quite a bit of detail about how they drilled into the amber, pumped out the blood from the mosquito stomachs, and extracted the DNA from the blood. Crichton assumes that dinosaurs, like birds, have nucleated red blood cells, making their blood a rich source of DNA. (Mammalian red blood cells lose their nuclei as they mature, so blood would not be an ideal source for harvesting mammalian DNA). And he assumes that whatever DNA they get will be torn and incomplete, so they will have to patch it together with snippets of DNA from living species.

He provided his chief cloner, Dr. Henry Wu, with a bank of sequencing machines that Craig Venter of Human Genome Project fame might have envied. However, sequencing machines generally need more

DNA than you would be likely to get out of a mosquito stomach that has been sitting in amber for millions of years. You would need to amplify the extracted DNA. Today, the most common method of amplifying DNA is the polymerase chain reaction (PCR). As its name suggests, PCR involves a reaction that feeds upon itself (like the atomic chain reaction that enables a nuclear bomb), and amplifies the DNA exponentially. Invented by Kary Mullis and a team at Cetus Corporation in the early 1980s, we can now use it to generate *lots* of DNA from even *one* snippet. And we have machines to carry out the tedious steps of heating and cooling the samples umpteen times. Sounds perfect for Henry Wu's needs!

But there is a catch. You have to know something about the sequence of the DNA before you can amplify it using PCR. PCR requires "primers" that bracket the stretch of DNA you are hoping to amplify (primers as in "priming the pump," not as in "undercoat of wall paint" or "book for beginning readers"). The primers have to match, at least approximately, the DNA sequence of the target you are hoping to amplify. So, you need to know something about your target. And Henry Wu knows nothing about the stuff he's pumped out of the mosquitoes. He cheerfully admits, "We won't know for sure, of course, until we extract whatever is in there, replicate it and test it" (99). And pointing to a table full of cloned eggs waiting to hatch he says, "That's a new batch of DNA. We don't know exactly what will grow out. The first time an extraction is done, we don't know for sure what the animal is" (107).

But let's step over this minor credibility gap. Wu also admits there are "gaps" in the dinosaur DNA. Some sequences are just too shredded up by all those millions of years in the mosquito's stomach to be recovered. He chooses to fill in the gaps with DNA sequences from living animals.

In several cases Dr. Wu chooses frog DNA sequences to fill in the gaps, possibly because even in 1990 a great deal was known about frog DNA. The African clawed frog, *Xenopus laevis*, is a common experimental animal, so more is known about frog genetics and embryology than about that of lizards, for example. However, one of the subplots in the novel originates from this choice. In order to keep control of the dinosaur population, the Jurassic Park cloners arrange for all of the animals to be female. Thus they would be unable to reproduce and their number would always be known exactly, enabling the staff to keep track of them. Unfortunately for these well-laid plans, amphibians may have the ability, under certain circumstances, to change their sex. By page 164, it has become clear that there are more dinosaurs than there should be, and suspicion grows that they must be making baby dinosaurs the way they did before they became extinct. As it turns out, the dinosaurs, which were designed to be exclusively female to prevent independent breeding, changed sex, made some baby dinosaurs the old-fashioned way, and managed to stow away on the supply boats and escape to the mainland.

To Henry Wu's credit, he had some concerns about the amount of tinkering he had to do in order to come up with a reasonably complete set of (mostly) dinosaur DNA. He was a little worried about the frog DNA he used to patch the gaps. (John Hammond ignored his concerns.) The problem is that sex determination in amphibians and reptiles is not completely under genetic control, as it is in birds and mammals. Mammals use the XX/XY system: the female is the default sex, the male is the heterogametic sex, and genes determining maleness are carried on the Y chromosome. Birds use the ZW/ZZ system: the male is the default sex, the female is the heterogametic sex, and genes determining femaleness are carried on the W chromosome. Amphibians' and reptiles' sex is influenced by their genes (some species use XY and some use ZW), but it is also influenced by the temperature at which the eggs are incubated. We don't know, of course, if dinosaurs shared this feature with modern reptiles, but if they did it could explain how some of the dinosaur hatchlings were not female after all.

As a further protection against the park dinosaurs escaping and repopulating the mainland, they were made "lysine dependent," meaning that they cannot synthesize the amino acid lysine and must have it provided in their food. The cells in our bodies, and those of dinosaurs, are primarily made up of proteins, which are manufactured by each cell in the body from components called amino acids. Herbivorous animals are generally able to synthesize all the amino acid building blocks, because most plant foods don't contain all of them. But many animals, in particular carnivorous ones, need to get some amino acids from their food. Humans, for example, have nine "essential amino acids"—meaning they are an essential part of our diet since we cannot synthesize them in our bodies. Meat products contain all the essential amino acids (histidine, isoleucine, leucine, lysine, methionine, phenylalanine, threonine, tryptophan, and valine). Vegetarians, however, must be sure to eat a combination of foods that will contain all of them, as most plant foods are deficient in one or another. The Jurassic Park dinosaur diet is supplemented with lysine, as they cannot synthesize it. The planners hoped this limitation would make them unable to survive off the island, but toward the end of the book there are reports of strange-looking lizards cruising across the countryside, eating mainly lysine-rich food. Presumably they ignore corn, which is deficient in lysine, and opt for beans, which are rich in it.

But even with all these high-tech aids, the fictional dinosaur cloners got dinosaur DNA, but no dinosaur nuclei. By the beginning of the twenty-first century non-fictional cloners had cloned frogs, mice, sheep, cows, horses, mules, cats, and a dog, but they did it using whole nuclei, not just naked DNA.

The fertilized egg (called a zygote) must not only divide again and again to form a ball of many cells, but must orchestrate the development of all the varied cell types that will eventually make up the newborn animal. Successful cloning, therefore, requires a nucleus from the donor to give up the simple life and resume the complex life of a zygotic nucleus. Scientists who make their living cloning things tear their hair out trying to devise better ways to "reprogram" adult nuclei so they can be successfully cloned into a waiting egg. This is why cloning is so hard!

And even if you did locate a complete dinosaur nucleus, what would you put it into? In cloning, the recipient egg is just as important as the donor nucleus, as it is the egg cytoplasm that provides the signals to reprogram the nucleus. It is the egg that instructs genes in the nucleus to make proteins needed for an embryo's development rather than whatever they were making when they were (for example) in a skin cell. It is the egg that instructs the donor nucleus to resume cell division.

In the world of mammalian cloning, eggs are the limiting factor, since it will likely take hundreds of eggs to get one successful clone. Even if there were no ethical qualms about human cloning, it would be challenging to get enough eggs. Getting hundreds of eggs from sheep is easier than getting hundreds of eggs from human donors. For one thing, you don't have to get permission from the sheep before harvesting their eggs. Jose Cibelli and his colleagues, who cloned the calves, collected their eggs from slaughtered cows, and then ripened them in a petri dish. Mark Westhusin and his colleagues, who cloned the first cat, collected their eggs from routine spaying procedures, and also ripened them in a petri dish before use.

It's a little harder to get human eggs in great numbers. After hormonal stimulation, an IVF clinic can usually retrieve ten to fifteen mature eggs from a human patient, either for her own use in *in vitro* fertilization or for egg donation. Egg retrieval is either minor surgery (laparoscopy) or an office procedure (transvaginal ultrasound). The hormonal stimulation has emotional side effects, and the retrieval process is uncomfortable and carries a small risk of complications, and so far it is not possible to collect immature human eggs and ripen them, as it is with cows and cats.

In this at least *Jurassic Park*'s dinosaur cloners had it easier. They used crocodile eggs with artificial shells.

The plot line of *Jurassic Park* makes use of one of the central facts of life, at both the individual level (embryonic development) and the population level (evolution): it is not predictable. Because they grow exponentially, both developing embryos and evolving populations are the kinds of systems that may not be successfully described by linear equations. They are often cited as examples by people who study Chaos Theory, or nonlinear dynamics, and are often described as "emergent" or "self-organizing" systems. Early on in *Jurassic Park* Ian Malcolm, expert on Chaos Theory, explains to Alan Grant, dinosaur expert, that the Jurassic Park project they are touring is doomed because the dinosaurs will be unpredictable. He provides readers with a mini-course on Chaos Theory, the branch of physics that attempts to account for the fact that many systems are not successfully described by the linear equations used in traditional physics.

The "chaos" in Chaos Theory doesn't mean messy or disorganized; it just means nonlinear and hence very difficult to predict. A common example is the weather, where tiny changes in atmospheric conditions in the Pacific Ocean can produce enormous changes in a weather system by the time it reaches New England. An example Malcolm describes is a pool table, where tiny variations in the air currents, bumps on the table top, and force of the shot make it impossible to predict where the ball will end up after ten ricochets off the table rim. Such systems are called "nonlinear." The equations that describe them contain a lot of exponents, and the graphs that map them contain curving lines instead of straight ones (hence the description "nonlinear"). This means that tiny changes in the initial conditions can lead to enormous variations in the final result, and seemingly simple systems can produce complex behavior. A simple mathematical demonstration of this is the fact that $2^{10} = 1064$, while $2.1^{10} = 3503$. A small increase in the number taken to the 10th power yields a product over three times larger. The upshot, as Malcolm repeats a number of times in the book, is that it is impossible to predict some very important features of the dinosaur park.

About halfway into the book, when a few problems have come up (the stegosaurs are sick; vermin are mysteriously disappearing; some of the dinosaurs are really, really ferocious), he fumes pompously, "I told him where the deviations would occur. Obviously the fitness of the animals to the environment was one area. This stegosaur is 100 million years old. It isn't adapted to our world. The air is different, the solar radiation is different, the land is different, the insects are different, the sounds are different, the vegetation is different. Everything is different. The oxygen content is decreased. This poor animal's like a human being at ten thousand feet altitude. Listen to him wheezing" (159).

"And the other areas?" asks Gennaro, the lawyer, also touring the park.

"Broadly speaking, the ability of the park to control the spread of life-forms. Because the history of evolution is that life escapes all barriers. Life breaks free. Life expands to new territories," Malcolm says ominously (159).

However, after taking the trouble to have Ian Malcolm educate his readers on Chaos Theory, Michael Crichton does not actually provide plot twists that are examples of mathematical Chaos Theory. The disasters that finally doom the park were certainly unexpected (by everyone except Ian Malcolm), but they were not unpredictable in the way that nonlinear systems are. John Hammond didn't listen to Henry Wu's concerns about the amount of genetic tinkering involved. He was also the victim of human greed and malevolence when his computer programmer disabled the security system (including the electric fences that enclosed the dinosaurs) so he could steal and sell ten dinosaur embryos to Hammond's competitor. The monitoring system that counted the dinosaurs was designed only to check whether any were missing, not whether there were too many. These events aren't nonlinear; they are just bad luck, human nature, and human error. They are examples of chaos in the popular sense rather than the mathematical sense. Ian Malcolm would have better spent his energy explaining Murphy's Law.

But even if the project weren't doomed to some kind or other of catastrophic failure, even a decade and a half later (*Jurassic Park* was published in 1990), we still can't resurrect dinosaurs. Michael Crichton did, however, resurrect Ian Malcolm, who apparently dies at the end of *Jurassic Park* of injuries inflicted by the rampaging T-Rex. He turns up in the sequel, *The Lost World*, claiming that reports of his death had been "greatly exaggerated." He is still a mathematician. However, in the second book his function as author's mouthpiece is not to explain Chaos Theory, but to discover that a prion disease will eventually kill off all the cloned dinosaurs on this second (actually first) island, and to finally explain why many of the dinosaurs are so ill-tempered.

The "Lost World" is a second island, which housed the actual research facility where the dinosaurs were cloned and the babies were reared to be taken to the park when old enough for release. At the start of the book, the island is long since deserted, but some of the cloned dinosaurs survive and have established a colony of sorts. This time the various protagonists gather on the island, attracted by irresistible curiosity over vague reports of possible dinosaur sightings in Central America. There are the usual thrilling adventures, and Ian Malcolm is again attacked by a rampaging T-Rex.

While drugged with morphine after the attack, Malcolm puts together the strange behavior of the dinosaurs with what he knows about the importance of child-rearing in transmitting cultural evolution. The cloned dinosaurs have no parents! Thus there has been no one to teach them how to behave and how to rear their young. The more primitive, such as Tyrannosaurus Rex, do this by instinct (as bottle-raised orphan kittens can grow up to raise normal kittens), but the Velociraptors, which are described in the novel as highly intelligent and capable of learning, need to be taught how to act. In the absence of adult raptors to transmit normal behavior, they are *terrible* parents and few of their young survive. Malcolm believes they will soon become extinct a second time.

As it turns out, the original cloners made a fatal error in designing their infant dinosaur formula. They used an extract made of ground-up sheep carcasses. If only they could have foreseen the dire consequences! In the late 1980s, when the fictional cloners were presumably at work rearing baby dinosaurs, no one had heard of "mad cow disease." By 1995, when *The Lost World* was published, Bovine Spongiform Encephalopathy, or "mad cow disease," was in the news all over the world. By 1996 it had become clear that the disease could spread to humans, and today British beef is banned from export to much of the world, while thousands of British cattle have been slaughtered in an effort to contain the disease.

Mad cow disease is now known to be caused by "prions." However, the causative agent and the mode of transmission remained a mystery for many years. Prions are not bacteria, viruses, or parasites. They are simply abnormally folded proteins, found primarily in the brain and nervous tissue of affected animals or people. They "reproduce" by converting normal proteins to their abnormal folding pattern. Thus experiments designed to discover the infectious agent failed time and again because the experimenters were looking for a virus, bacterium, or parasite. Stanley Prusiner won the 1997 Nobel Prize in Physiology or Medicine for establishing that the spongiform encephalopathies are caused by prions.

Prion diseases have been known for decades, as scrapie in sheep, kuru among New Guinea tribesmen who practiced cannibalism, and Creutzfeldt-Jakob Disease in Europe. The disease is always fatal, and involves a gradual breakdown of neurological structure and function. They are called "spongiform encephalopathies" because the diseased brain becomes riddled with holes like a sponge. A minority of Creutzfeldt-Jakob cases are genetic, but most are acquired during surgery, such as in the case of a corneal transplant from an infected individual. In sheep, scrapie can be transmitted from mother to fetus, and also horizontally among the members of a flock, so it has been impossible to eradicate. Among the Fore tribesmen in New Guinea, kuru was spread by eating the brains of infected people. (The Fore have since given up the practice of ritual cannibalism and the disease has died out.) BSE was apparently passed from sheep to cattle via the modern practice of using cattle feed that included ground-up sheep parts (these concentrated food pellets help fatten them up for market). Among cattle it has not been shown to be transmissible by any other means than eating infected sheep parts, but some people who have eaten infected cattle have developed a neurological disorder similar to Creutzfeldt-Jakob disease.

The dinosaurs of the Lost World appear to be dying off from a prion disease, acquired from their infant formula that included extract of whole sheep carcasses, some of which must have been infected with scrapie. It is spread among the predators and scavengers by eating infected carcasses and among some other species by their habit of eating the poorly digested droppings of other dinosaurs.

But if the cloned dinosaurs of Jurassic Park weren't doomed by Chaos, prions, and bad parenting, they may have been doomed by their status as clones. As Dolly and the cloned mice and calves got older, it appeared many of them were not completely normal. Dolly was fat and had arthritis. She was euthanized at the age of six (young for a sheep) because of her arthritis and persistent respiratory infections. Obesity is now known to be common among cloned mice, many of whom also die young. An article in *Science* optimistically entitled "Cloned Cattle Can Be Healthy and Normal" actually contains some sobering data on how many cloned calves died before or shortly after birth. It is not clear whether these abnormalities are due to incomplete reprogramming of the donor nuclei, to incompatibility between the donor nucleus and the recipient egg, or to some other problem. But at this juncture cloning by nuclear transplantation, even of living donors into eggs of the same species, is a very uncertain proposition.

It's hard for a scientist to feel friendly toward Michael Crichton. The guy seems really hostile to science and to many of the people who practice it. Although the dinosaur cloning experiment never would have happened without the tycoon Hammond paying for it, the real focus here is that cloned dinosaurs ran amok, and the scientists who cloned them sold their services to the highest bidder without considering whether the project was advisable or not. Ian Malcom, the chaotician, is Crichton's alter ego here. Shot full of morphine because his leg has been mangled by a T-Rex, Malcolm rambles and fumes at some length toward the end of *Jurassic Park*. "Discovery is always a rape of the natural world. Always" (284). "Scientific power is like inherited wealth—attained without discipline.... Old scientists are ignored. There is no humility before nature" (306). As described by Malcolm, science is inherently unable to make any decisions about how to use the knowledge it may uncover: "Science can make a nuclear reactor, but it cannot tell us not to build it. Science can make a pesticide but cannot tell us not to use it. And our world starts to seem polluted in fundamental ways—air, and water, and land—because of ungovernable science" (312). His drugged state allows him to rant like this despite the fact that he is in some ways one of the scientists he so disparages. He plows on and on, and apparently dies at the end, although he is resurrected in *The Lost World* to serve again as the author's voice.

In *The Lost World*, the anti-science grumbling is done by Marty Gutierrez, a field biologist in Costa Rica. He remembers a time when scientists had "a passion to learn about something for its own sake," in contrast to the current "looter mentality" in which "people aren't studying the natural world any more, they're mining it" (31–32).

No scientist likes to read this kind of thing in a bestselling novel, but Crichton brings up some issues that need to be addressed. How should science be financed? Who should decide how or whether scientific discoveries should be used? Should it be available for profit or financed by the taxpayers for the public good? When John Hammond is recruiting Henry Wu to his scientific team, he tells him not to take a job at a university if he really wants to do great science: "Universities are no longer the intellectual centers of the country. The very idea is preposterous. Universities are the backwater" (123). Hammond doesn't spell it out, but the implication is that universities are the backwater because money is flowing elsewhere.

Interestingly, the real villain in both the novels is Lewis Dodgson, a geneticist/industrial spy for BioSys Corporation. He is the one who bribes the computer tech at Jurassic Park to disable the security systems and steal dinosaur embryos for him. He is the one who tracks down the Lost World in a second attempt to steal the cloning technology. (He is eaten by baby Tyrannosaurs at the end of *The Lost World*). It is possible that Crichton's real beef is with the commercialization of science, not with science itself. He puts bitter words into Ian Malcolm's mouth, but none of the "real" scientists are portrayed as jerks.

It is not clear whether it is Crichton's intention, but he makes a good case for increased government funding—and therefore oversight—of science. We cannot exactly forbid private companies from doing biological research (forbidding it in the U.S. would just drive them offshore), but greater support from the National Institutes of Health and the National Science Foundation would put the universities and research institutes clearly in the mainstream, not the backwater. It might mean less mining of the natural world and more study of it.

SANDY BECKER has been a practicing scientist for twenty-seven years. The first twenty-five years were spent doing research in developmental biology at Wesleyan University in Connecticut. To supplement her income she moonlighted as a science journalist. Before discovering her true calling as a biologist she worked as a writer of civil service tests, a fifth-grade teacher, a folk singer, and a mom. Since leaving Wesleyan she has worked for Advanced Cell Technology, a biotech company in Massachusetts, hoping to make something medically useful out of embryonic stem cells.

CRICHTON TRAVELS IN TIME

Joel N. Shurkin

Timeline *is both a techno-thriller and a historical drama. Although ostensibly about an academic time travel expedition/rescue mission back to fourteenth-century France—where we are immediately reminded that life then was nasty, brutish, and short—we find out that the travelers in* Timeline *are, in fact, actually passing into entirely separate, yet similar, universes. Isn't this "out there," even for Michael Crichton? Joel N. Shurkin fills us in.*

A History of History

Writers find time travel irresistible. In his novel *Timeline,* Michael Crichton entered a field rich with some of the most inventive and clever of writers in a tradition going back 300 years.

While his book was a bestseller and was turned into a movie, it is not regarded as one of his best. It does, however, show him at what he does best: taking current scientific thinking and extrapolating and staying within the bounds of what we know until the time when his imagination requires the existential leap he needs for his plot.

Time travel stories generally run into two literary genres: fantasy and science fiction. Fantasy writers usually don't bother with explaining how time travel is possible, while science fiction writers do what Crichton does: confront the science and then ignore what needs to be ignored and take it from there. That's the *science* in science fiction.

The first example of time travel in English literature may be *Memoirs of the Twentieth Century*, written in the eighteenth century (1733) by Samuel Madden, in which documents from 1997 and 1998 showed up in the story. The documents originated from a guardian angel fantasy as opposed to science fiction. There probably was not enough known science to do science fiction in 1733. Other examples appeared in the early nineteenth century, including a short story in the *Dublin Literary Magazine* in 1838 in which a traveler went back in time and met the Venerable Bede (seventh century), but like Rip Van Winkle, it could have just been a dream. Scrooge went back and forward in time in Dickens's *A Christmas Carol* (1843), but again, that may just be a delusion. On the other hand, Scrooge was able to change the future. Mark Twain's *A Connecticut Yankee in King Arthur's Court* (1889) is a famous example and one of the first in which a character changes history. That won't do, of course, and all his work had to be undone at the end. In Edward Page Mitchell's *Clock That Went Backward* (1881), two boys found an old clock which transported them to sixteenth-century Holland when they wound it backward. The most famous nineteenth-century example, of course, is H. G. Wells's *The Time Machine* (1895), in which someone invented a vehicle for the sole purpose of traveling in time. It looked surprisingly like a bicycle.

Many stories have people on one-way trips, like Twain's hero. Jack London wrote about one in *Before Adam*, and H. Rider Haggard did it in *The Ancient Allan,* in which his white hunter, Allan Quatermain, was thrown into the head of a prehistoric man. People going back in an attempt to change history is a popular theme. In Robert Silverberg's "The Assassin," someone tried to prevent Lincoln's assassination, and in Maurice Vaisberg's "The Sun Stood Still," someone tried to kill Joshua at the Battle of Jericho. Sometimes the history we know is described as the result of change, for instance in Harry Harrison's *The Technicolor Time Machine,* a movie producer trying to make a movie about the Viking landings in North America went back in time, but when the Vikings didn't show up he had to import them, changing history.

There was a time police squad in *The Legion of Time* (Jack Williamson) to prevent historic alterations, and time patrolmen in *The Guardians of Time* (Poul Anderson). In Nathan Schachner's "Ancestral Voices," a man went back to kill his grandfather. Much more about that later.

Crichton completed *Timeline* in 1999 and like all his novels, it reads as if it could be turned instantly into a screenplay, as indeed many of them, including this one, have.[1]

[1] The screenplay, written by others, kept the violence but threw out the imagination and what little wit resided in the novel, and the movie received uniformly bad reviews. Unusual for him, Crichton gave away the rights to the book because he apparently was caught in political contretemps between a new agent and the Hollywood establishment, and no one would buy the rights.

As with all his books, *Timeline* was based on solid science from which Crichton took imaginative flight, and the dialog is full of references, true and fictitious, giving it all the look of real scientific scholarship. You have to Google extensively to figure out which are real and which Crichton made up. He is very good at this. Crichton, a medical doctor, couldn't resist a little medicine in the plot but mostly he was writing about quantum time travel and the fourteenth century, and one presumes he had a wonderful time researching both. He has five pages of bibliography—mostly books on medieval France, including a medieval cookbook—and his description of Dordogne and Castlegard in southwest France were likely quite accurate (Dordogne is real, in Aquitaine; Castlegard not, but it is likely a good facsimile of a real fortress of the time). One also suspects that he spent time researching in France (the descriptions were too good) and that's what tax-deductible business trips are for. His history contained at least one flub: a modern character introduced the local combatants to "Greek fire," the medieval equivalent of napalm. In fact, Greek fire goes back to, well, a Greek in the Byzantine Empire, and was used in the seventh-century siege of Constantinople. It is unlikely that the military in the fourteenth century had never heard of it. He also wrote that the modern concept of the Dark Ages where life was indeed short, brutish, and decidedly unchivalrous was a Renaissance invention. You would not know it the way his knights behaved.

He had some hilarious dialog, an attempt to sound archaic.

> "My Lady... your public manner and sharp discourtesy provoke many to laugh behind my back, and talk of my unmanliness, that I should tolerate such abuse.
>
> "It must be so," she said. "For both our sakes. This you know full well."
>
> "Yet I would you were not quite so strong in your manner."
>
> "Oh so? And how then? Would you chance the fortune we both desire?" (*Timeline* 282)

Did they really talk that way?

The Plot

The story begins with the appearance of a mortally injured man in the Arizona desert. He is raving and suffering from strange injuries. The weaving of his clothing is odd—disconnected as if woven from multiple pieces. He seems discordant, oddly and physically disunited. Indeed, some of his blood vessels are misconnected and his fingertips, cut off from his blood supply, are falling off. (In the film, the writers even had his cardiovascular system misaligned, which would have killed him instantly and probably gave Dr. Crichton heartburn). He turns out to be a noted physicist who was reported missing from a New Mexico corporate lab run by International Technology Corp (ITC) which mysteriously researches commercialization of quantum physics. Robert Doniger, an arrogant, brilliant, driven, and unpleasant scientist, runs ITC. He is the type of scientist found in virtually every Crichton book. Fiction requires conflict, and these characters, who by no means are absent from the real world of science, are perfect literary foils. ITC, the stereotypical evil multinational, has developed a method of time travel working with the latest research.

Meanwhile, a group of Yale archeologists are digging around the site of the fourteenth-century fortress and monastery in Dordogne, funded by ITC, the scene of a (fictitious) English defeat in the Hundred Years' War. The battle was between the forces of Sir Arnaut de Cervole, a mercenary, and Sir Oliver de Vannes, an English knight. De Cervole was a real person, known as the Archpriest. Vannes is apparently fictional although there is a town by that name in Brittany. The company wants the archeologists to recreate Castlegard as a theme park (shades of the dinosaur cloning in *Jurassic Park*) and is funding the work so that it can claim that the replica citadel and town were absolutely perfect. Or so it says. When the lawyer representing ITC mentions places and buildings that she could not possibly know about, things the archeologists haven't yet excavated, the chief scientist, Edward Johnston, grows suspicious and insists on speaking to Doniger. He flies with the lawyer to New Mexico and promptly disappears. Sometime later, the young scholars working with Johnston discover a set of modern bifocals in the ruins, and a manuscript containing a plea from Johnston for help which was written 700 years ago.

ITC has found a way to transport people and information back through time and they sent several, including Johnston, to 1357 Castlegard. But there are problems with the technology. The physicist found in the desert was one of the victims. There are transcription errors that build with multiple journeys. Johnston is trapped back then.

Three of the four young scientists are sent back to 1357 to rescue Johnston. At this point the book falls apart, turning into a conventional chase against time with Crichton saving his characters with excessive good luck, miracles, and fortuitous arrivals of *deus ex machina*, occasionally breathtaking and eventually tiresome. No one appears to break a bone—you either die or you don't. Meanwhile in the plot, disaster strikes back in twentieth-century New Mexico when all the time travel machinery gets wrecked. The newest voyagers, apparently, are trapped in the fourteenth, too, unless the people at ITC and their remaining colleague, working against Doniger's orders, figures out a way to bring them back.

The time clock, counted off in subtitles, is clicking. Needless to say, the professor gets rescued and two of the three young scholars return safely just at the last possible second. One remains behind, marries a promiscuous princess, and lives a happy, short life back then—a time he always admired. Unlike many Crichton novels where he has problems working out a satisfactory ending, his characters in *Timeline* gets revenge against Doniger, but it would be callous to describe it. Think rats and fleas.

The problem for novices is knowing when the science really ends and the fiction begins. Crichton is a master at fudging the lines.

Time Travel

Any explanation of time travel runs into two concepts. The first is the Newtonian and Einsteinian view of determinism and causality, a world where if A bumps into B, B can bounce into C in that order. B doesn't move until A hits it. C does not move before the collision, we are very sure. We have laws to explain what happens, including inertia and conservation of energy and Newton's third law of motion. If B moves, we know why: it was hit and is reacting to the force of the hit. Then there is the quantum physics concept, which Einstein loathed, in which the world is not deterministic; nothing happens necessarily in sequence and what is normally thought of as logical ain't necessarily so. Indeed, adherents argued you cannot even discuss quantum events in normal language. In the quantum universe, existence is an infinite collection of random probabilities and nothing more.

Time travel rams right into the Newtonian laws of classical physics. For instance, if a time machine suddenly appears in the past, does its sudden appearance violate the law of conservation of matter? In the classical view, it would seem to depend on when and where it appears. If you send information back in time, that requires the expense of some energy, so does that violate conservation of energy? Even the laws of motion would seem violated in Newtonian physics. The violations may not apply, however, in the quantum reality.

Neither classic nor quantum physics recognizes past or future. There is no *here* or *now*. The physical world is too random for that. Nothing prevents a broken glass from reconstituting itself. It is just not likely to happen. In the quantum universe, something can be in two places at the same time or the same place in different times. Unfortunately for those embedded in deterministic logic, quantum physics has been proven repeatedly; reality apparently is a collection of probabilities.[2]

The physicist Stephen Hawking suggested that since we have never run into time travelers in our lifetime, it is logical to conclude humans never invented time travel. But that's another problem. Maybe they are hiding. It's not that we haven't tried to locate them. On May 7, 2005, the Massachusetts Institute of Technology (MIT) held a Time Traveler Convention, inviting any time travelers to show up for lectures and refreshments. The specific time, advertised and posted continually, is (or was) 0200 UTC at 42.360007 degrees north latitude, 71.087870 west longitude on the MIT campus.

The literary concept breaks down into two modes. In the most common, the author finds a way to produce a device that somehow travels backward or forward through the time-space continuum requiring complicated, sometimes not logically satisfactory, explanations for how it is done. See Wells's bicycle in which the time traveler sees the future in fast-forward.

---

[2] Crichton has quotes from real scientists pointing out, as Richard Feynman says, "Nobody understands quantum theory."

The other path, which requires at least as great a jump in disbelief, is to assume the travel is not so much through time as through dimensions to an alternate or parallel universe working in a somewhat out-of-sync time frame. Science fiction writer Larry Niven in his classic essay "The Theory and Practice of Time Travel," called it "sideways-in-time travel," or travel between multiple time tracks. The traveler goes to another dimension or universe in which the events he is trying to reach are happening now, not 700 years ago in this timeline. Niven said he hates those stories because "they're too easy to write. You don't need a brain to write alternate-world stories. You need a good history text." Crichton, however, never takes the easy way out and most certainly uses his brain and good historical research.

Time travel novels ought to have some explanation, although some fantasy authors don't bother. In the recent bestseller, *The Time Traveler's Wife*, Audrey Niffenegger moved the protagonist back and forth through a woman's life without any attempt to explain how it happened. But it isn't important to the plot, which was mostly metaphor anyway. Crichton, ever the scientist, could not possibly accept such a cop-out and neither would his readers.

Crichton adapted the parallel universe model. He devised a way to fax his characters to fourteenth-century France.

There are scientists who think that alternate universes represent the only way that time travel to the past could ever happen. As with almost every science fiction novel, the author has to set up a scene to describe the world to the reader, usually by having one character explain it to another. That is often clumsy because the fictional listeners probably should have known it all ("As you know, Ted, Kaufman invented time travel in 2012 and. . . ."), but Crichton is a master at this trick and was at his best in *Timeline*. The young scholars who are about to be shipped to the fourteenth century have to understand what is going to happen to them. They are not physicists but liberal arts majors or archeologists who do not understand quantum physics—just like the reader. The fictional explainer is an engineer at ITC named Gordon.

Crichton is a master at name-dropping and the razzle-dazzle is impressive. He starts with the difficult task of explaining quantum physics, quanta, Planck, Einstein, and everyone else. He calls it "the most proven theory in the history of science," which is a stretch. He even starts with the Big Bang. The underlying scaffolding for his science, the quantum world, is nothing like the one we see and measure. Things that can't happen here are normal there. One must suspend belief.

Much of the basis for Crichton's technology comes from physicist Kip Thorne's 1994 book *Black Holes and Time Warps*. Crichton credited the parallel universe concept correctly to Hugh Everett III, a student of John Wheeler's at Princeton, and his "relative state formulation" or Many-Worlds Theory developed in 1955, a theory almost universally rejected by his colleagues then, but crucial to the device in *Timeline*. The Many-Worlds Theory makes the simple conclusion that one probabilistic outcome is as real as any other. He predicted an almost infinite supply of alternate universes branching away from each moment of now. Each decision we make starts another universe with another timeline. He posited an infinite number of copies identical to our present, but there is always something slightly askew in each one.[3] There would be in Everett's cosmos an infinite number of universes, including a timeline in which Hitler wins the war, as terrifying a concept as that might be, the Chicago Cubs win the World Series, or an aspiring young writer falls into the Avon river in the sixteenth century and drowns. Every act in our universe is mirrored in another, but not at the same time or in exactly the same way.

The Twin Slit Experiment

As with much of Crichton's science fiction, he builds from real experiments and real theories. He has Gordon describe the twin slit experiment, a demonstration beloved by first-year physics students. If you shine a light through one slit in a piece of cardboard at another cardboard behind it, it produces one bright slit-like line on a screen. Two slits produce not two lines on the screen but a pattern. The light goes to some places and not to others. Then things get spooky. Four slits produce not more lines on the screen but half the number, only two. How is that possible? The reason given (the "nineteenth century reason," the character says) is that light is a wave, and what you are seeing is one wave interfering with another. But since Einstein proved

[3] Everett gave up trying to convince his colleagues it was worth looking at and finally quit physics in disgust, becoming a multimillionaire businessman.

that light also was a particle, the character says that answer can't be right. Yes, it can.

A twenty-first-century quantum physics answer is that the particles interfere with each other—quantum interference. The denser areas of probability in the interference pattern represent the more probable phenomena, while the thin areas represent the least probable.

But it gets wilder. If you manage to produce a beam of light over a course of time that consists of a stream of single photons in a line, you still get a discernible pattern. The light lands in some places, but not in others. Since there are no other photons involved, what could be interfering? For the purposes of the plot, Crichton's answer is an alternative universe, a cosmos of multiverses. As the experimenter shines the single photon beam at the screen, another experimenter in another universe is doing exactly the same thing. The photons in our universe are being interfered with by the photons in the other universe. We can't see it. We can't measure it. But the photons in our universe are running into photons in the other and getting bumped aside.

> "It's the nature of the multiverse," Gordon said. "Remember, within the multiverse, the universes are constantly splitting, which means that many other universes are very similar to ours. And it is the similar ones that interact. Each time we make a beam of light in our universe, beams of light are simultaneously made in many similar universes, and the photons from those other universes interfere with the photons in our universe and produce the pattern that we see....
>
> "And that proves that the other universes exist" (*Timeline* 130).

Well, it doesn't, of course. The generally accepted explanation (as Crichton undoubtedly knows) is the duality of light: light is a stream of particles that can behave like waves. In the slit experiment, the photons are interfering with each other. If this sounds illogical, well, so is quantum physics and those are the kinds of answers you get. Indeed, as Crichton also probably knows, this idea is getting increased attention in the physics community.

Everything that happens in the universe, Everett added, produces another universe in which something else happens and the number of possibilities is limitless. According to physicist Thomas Hertog, "Quantum mechanics forbids a single history."

This would be a good time to bring up Schödinger's Cat, the hilarious thought experiment that best exemplifies the strangeness of quantum theory. Erwin Schrödinger who, like Einstein, hated quantum theory, tried to prove the absurdity of it all:

Put a cat in a box. Add a radioactive atom with a five-minute half-life—a 50 percent chance of decaying within five minutes. If it decays within five minutes it triggers the breaking of a vial containing poison gas that would kill the cat. If it doesn't, the cat comes out of the box angry but unscathed. The odds for the cat's survival are fifty-fifty. Wait five minutes and open the box. Is the cat dead or alive? It obviously has to be one or the other, Schrödinger insisted.

Nope, said the quantum theorists of the Copenhagen school. Until the box is open, the cat exists in an intermediate state, neither dead nor alive. When we open the box the cat settles on one or the other, dead or alive.

Crichton would be sending his characters from the world in which the cat is neither dead nor alive, getting there before the vial breaks.

For the next stage, Crichton again relied on jargon: wormholes and quantum foam. "We make wormhole connections in quantum foam," the engineer says.

> "You mean Wheeler foam?" [asks an incredulous anthropologist] "Subatomic fluctuations of space-time?"
>
> "Yes."
>
> "But that's impossible" (*Timeline* 131).

Wheeler at the Institute for Advanced Study in Princeton is responsible for much of this. Wormholes are theoretical shortcuts through space and time originally described by the German physicist Hermann Weyl in 1921 and named by Wheeler in 1957. The name comes from this analogy: think of a worm on the skin of an apple. If the worm wants to get to the other side of the apple he can either crawl all the way around to the desired point or he can burrow his way through the core to get to the other side, cutting through the diameter rather than going around the circumference. He digs into his wormhole. In Wheeler's analogy, the apple is space-time and a wormhole is a shortcut through space-time. In theory, if you used a wormhole you could get to where you wanted to go much faster than it would take you to make the trip at the speed of light. Niven and Jerry Pournelle used it in *The Mote in God's Eye*, a classic sci-fi novel. It appears throughout *Star Trek*. Wormholes are how the protagonists in the television program *Stargate* get to other worlds.

Wormholes might permit this kind of travel because you never actually exceed the speed of light going through the wormhole. They are permitted in general relativity.

The other piece of jargon, quantum foam, also comes from Wheeler. This is the ripples and bubbles of space-time. Quantum foam is what space-time looks like at very close range. It is based in the Uncertainty Principle of Werner Heisenberg, better known to physicists as the Principle of Interdeterminacy, which specifies how empty a region of space can be. Particles and antiparticles, with both mass and gravity, are constantly popping into and out of existence, always changing the quantum landscape. In this concept, gravity has the same fundamental properties as the other forces in nature, such as gravity and the weak force of radioactive decay. It is therefore described by quantum mechanics. In Heisenberg's theory, you cannot know both the geometry of space-time or how that geometry is changing. Wheeler said that if this were so, space-time would appear foamy. Black holes would appear in the foam and then disappear quickly. They can do so without violating the law of the conservation of matter. At this point, some physicists think alternate universes ("baby universes") also would appear. Here is the core of Crichton's technology. Michael Crichton never makes up science; he extrapolates from it.

Paradox

Time travel runs into two paradoxes that, if nothing else, would make the subject great fun. One is the Twin Paradox, the other the Grandfather Paradox. Then there is the Butterfly Effect.

The Twin Paradox (from the special theory of relativity) is often called space contraction or time dilation. It holds that an object shortens along the direction of its motion relative to an external observer. The faster it goes, the shorter it becomes to an external observer. You won't notice it until you start approaching the speed of light. Time also shortens since space and time are facets of the same thing. Clocks slow the faster you go, a phenomenon now proven in space experiments.

Take two twins. One is an astronaut. If the astronaut leaves for a distant star at nearly the speed of light, time slows for him relative to time observed back on Earth. Time on Earth speeds up relative to time on the starship and would appear to run slower as seen from the astronaut's perspective, while an Earth observer would see time on the spaceship going in fast-forward. Should the astronaut return, he will find himself in the future meeting his great-grandchildren or beyond. Years, perhaps centuries, will have gone by. His twin would have long since died. For both the astronaut and the folks back home, time has passed normally and everyone is born, ages, and dies normally, but the returning astronaut has arrived in *his* future, *their* present.

Most fun of all is the Grandfather Paradox, which goes back to the French science fiction writer René Barjavel in 1943. What if you could travel back in time and kill your grandfather before he met your grandmother? That would mean you were never born, which means you could not go back in time to kill your grandfather. The only logical conclusion from that paradox is that time travel is therefore impossible.

One way of getting around this is to put your foot down and deny the paradox. It's not time travel that is impossible, it's the paradoxes. Since you exist, you could not possibly have killed your grandfather. Your trip must be complementary with the state from which you left. Anything you do in the past has been part of history all along and the time traveler can't undo that because it would create an inconsistency. Too bad. Think of going back and strangling Mrs. Hitler's little baby boy.

For the Butterfly Effect, we can credit Edward Lorenz (1963) who postulated that the flapping of a butterfly's wing could lead to a cascade of events that ended up producing a tornado. His notion is an offshoot of Chaos Theory. In time travel, the effects of any one incident, no matter how minor, could have rippling effects that could dramatically change history. In Ray Bradbury's 1952 short story, "A Sound of Thunder," a dinosaur hunter stepped on a butterfly, leading to a series of events that changes everything from the spelling of English words to an election. Homer Simpson went back to the age of the dinosaurs and killed the beasts off with a sneeze, producing a future in which his children were well-behaved and the sky rained donuts. It's the domino effect writ large.

Crichton rejects the Butterfly Effect and evades both paradoxes, finessing around the first by using parallel universes instead of true time travel and simply ignoring the second.

Doniger said that a single individual really cannot change history. As an example, Crichton's character said, you go to a Mets-Yankees baseball game. The Yankees won the game in history but you want to change the outcome. What do you do? You can't go out onto the field, the chances of you sniping the pitcher with a rifle are limited, and gassing the occupants at Yankee stadium is virtually impossible. The Yankees will always win. "You remain what you always were, a spectator." It's not a great example because in this situation, you are too removed from the probability. As for killing your grandfather, Doniger said too many things can go wrong that will prevent you from carrying out the task and running into the Grandfather Paradox. You may not actually meet him. You could be run over by a bus on the way, get arrested by police, get there at the wrong time—in other words, life is much too complicated for anyone to go back and alter history. Forget butterfly's wings.

This isn't true, of course. Individuals change history with single acts all the time. Think of Lee Harvey Oswald.

In *Timeline*, Crichton had his protagonists killing off dozens of people, which would have far more potent results to history than the mere fluttering of an insect. Every one of those slain would have potentially produced children who now would not be born, wiping out whole family lines; people who might have taken actions that had great ramifications and consumed or produced things that affected history. One character even remained behind and produced children that would not have been born but for the time intervention. Crichton simply ignored the possibility of the laws of unintended consequences, as his heroes crashed their way through fourteenth-century France.

Crichton, however, didn't make this up. In the mid-1980s, Igor Novikov stated that if an event exists that would give rise to a paradox or to any change in the past whatsoever, then that event can't happen. In other words, there can be no paradoxes. His assertion is known as the Novikov Self-consistency Principle. He didn't like the Grandfather Paradox as an example so he invented the billiard ball metaphor. Fire a billiard ball into a wormhole so that it would go back in time and collide with its earlier self. Too many things would happen to prevent that, including the later ball not hitting the earlier ball squarely, Novikov said. Indeed, he argued, the probability of the ball coming back and hitting its former self perfectly was zero. That's what Gordon was saying in the novel. Nothing a time traveler could do would change history. (Philosopher Paul Horwich added that you could not go back and kill your baby self, called auto-infanticide by philosophers. You were obviously already alive.)

This, incidentally, implies there are limits to free will. You will do what you have done whether you like it or not. See *Harry Potter and the Prisoner of Azkaban* for another example.

This deterministic view of history also has a modifying element. You might be able to change history in any point in time, but history will heal itself and things will end up the same. As Niven wrote, "probabilities change to protect history."

Crichton accepted Niven's Law: "If the universe of discourse permits the possibility of time travel, and of changing the past, then no time machine will be invented in that universe." To have time travel, you have to preclude the possibility of changing history. If you changed history even a little bit, you would create a universe slightly different than the original, and the next time traveler would do the same. Eventually, Niven postulated, you would create a universe in which no one invented time travel. You can stop right there.

The Technology

The analogy used to describe ITC's technology is the fax machine. You can't send a piece of paper with writing through a telephone line, but you can break it down into digital information and transmit that to have it reconstructed on the other end. With a clear line and a really good fax machine you would be hard pressed to tell the difference. Anything can be transmitted, Gordon tells the scholars, as long as it can be captured digitally, compressed, and encoded. But doing that with humans would seem impossible. A living human contains far too much information. Not so, says Gordon. Dropping more jargon, Crichton has him explain that the compression and coding made use of a "lossless fractal algorithm."

There is such a thing, going back to a paper published by Michael Burrows and David Wheeler in 1994, originally designed for text but later adopted for transmitting image information, including fractals (a repeated geometric pattern made up of parts, each identical to the other but smaller). Because they repeat, referring to iterations rather than repeating them can compress fractals. There is no loss of data. As Gordon describes it, suppose you have a picture of a rose made up of a million pixels. Each has a location (up and across) and color, so each pixel would be described by three numbers—3 million numbers. But many of those pixels are just red surrounded by other red pixels. The coding tells the computer to make that pixel red and make the next ten or 100 red as well without having to actually describe each of the 3 million. Then it can switch to another color. The information which incorporates only the unique attributes of the pixels is not the picture, but instructions for reconstructing the picture.

But, as one of the archeologists pointed out, that's just a two-dimensional picture of a rose. Now we're talking about a breathing sentient human being and that is a different matter. Gordon claims it is only different in the amount of information contained in the human and in order to handle that, one needs parallel computing, tying a batch of microprocessors together. ITC claims to have harnessed 32 billion! Impossible? Ah, not with a quantum computer, Gordon says.

Crichton credits Richard Feynman with the concept. Feynman theorized in 1981 that it might be possible to make use of quantum states to code information. In a regular microprocessor, computing is done with ones and zeros; bits representing on or off states. It is one or the other. In a quantum computer you could have a zero or a one or both a zero and a one (superposition) and a numeric representation of the probability of it being a zero or a one. There being thirty-two quantum states in an electron, that means you can get thirty-two times as much information in each piece of information. This makes no sense in the classic world, but we're talking quantum theory here. Quantum computers don't count bits; they count *qubits*. People are actually working on this.

Even one of those machines would fail to capture all the "information" in a living human, so ITC uses parallel quantum computers (microprocessors linked together) and that conceit allows Crichton to make a giant leap. It is difficult to make predictions in the technology world, suffice to say it will be decades before anyone gets that kind of computing power, but this is fiction, of course. To let Gordon describe it:

> "[Y]ou can indeed describe and compress a three-dimensional living object into an electron stream. Exactly like a fax. You can then transmit the electron stream through a quantum foam wormhole and reconstruct it in another universe. And that's what we do. It's not quantum teleportation. It's not particle entanglement [more jargon]. It's direct transmission to another universe" (*Timeline* 138).

Technically he *is* describing teleportation, as in "beam me up, Scotty." In this instance, the teleportation was to another universe with a slightly askew timeline.

ITC's time travelers first go into a device like an MRI—an "advanced resonance imager." Again, there is such a thing, but hardly what Crichton is describing. He doesn't give a lot of detail. Gordon describes it as using "superconducting quantum interference devices," SQUIDS. He's right on here. SQUIDS are used to measure very small magnetic fields. The traveler stands naked in a tube while the machine "calibrates." It hums, whirs, and lights flash. It takes only a few minutes to image the body. The computer doing the calculating does it in zero time. This is the quantum world, after all. The computer is doing what your fax machine does—storing information electronically before it is transmitted, greatly speeding up the process.

The actual transmission is done in telephone booth–like devices in a room shielded from outside interference by walls of water. The data from the first scanning is combined with the real-time scanning to compensate for any changes and verify the veracity of the information. Gordon explains it as Closed Timelike Curve technology (CTC), which incorporates the iridium-gallium-arsenide quantum memory, which Crichton avoids explaining. CTC's have been theorized as a closed world line. It keeps snapping back on itself, which is crucial to the plot and incidentally, if the theory is true, means time travel is possible.[4] The cylinder makes use of lasers, niobium, and polymers, all nouns that enhance verisimilitude. The machine is so sensitive in the quantum universe it is capable of detecting the arrival of a time traveler before the traveler leaves, Gordon says. In the classic worldview, cause comes after effect. In the quantum world, effect can be simultaneous to cause or even precede it.

The travelers get compressed, like their data, until poof, they disappear.

They remain conscious throughout the ordeal and it is "not an unpleasant experience," the scholars are assured. They need the assurance as they watch a woman and her cylinder shrink and vanish.

Now comes the hard part. When you fax a letter, the letter stays on your fax machine and a facsimile appears elsewhere. That's not quite how this technology works.

> "But this, uh, method of shrinking a person, it requires you to break her down—"
>
> "No. We destroy her," Gordon said bluntly. "You have to destroy the original, so that it can be reconstructed at the other end. You can't have one without the other."
>
> "So, she actually died?"
>
> "I wouldn't say that, no. You see—"
>
> "But if you destroy the person at one end... don't they die?"
>
> Gordon sighed. "It's difficult to think of this in traditional terms," he said. "Since you're instantaneously reconstructed at the very moment you are destroyed, how can you be said to have died? You haven't died. You've just moved somewhere else" (*Timeline* 157).

This is not entirely off the wall. In 1993, a group of scientists including Charles Bennett of IBM showed that perfect teleportation is possible but *only if the original is destroyed*—exactly what Crichton was doing here. There is no reference to the Bennett paper in his bibliography, so it may just be a lucky guess—or an incomplete bibliography. It violated no known law of science, although it seemed at first blush to violate the Uncertainty Principle, which argues that the closer you investigate something, the less information you can glean from it,

[4] At least one Nobel laureate, David Politzer, has suggested this is true.

hence, a perfect copy is impossible. And there is the conservation of matter and energy, but that may be rhetorical in the quantum world. According to the Einstein-Podolsky-Rosen effect, a paradoxical feature of quantum mechanics (measurement performed on one part of a quantum system can have an instantaneous effect on the result of a measurement performed on another part), you can actually send some kinds of information that can be delivered backward in time. In other words, it arrives before it is sent.

Gordon says the woman is in another universe. She's been faxed there. He is asked, if that is so, then there must be a receiver on the other end. No, he says, no receiver is needed, "because she's already there."

At the moment of transmission, the person is already in the other universe. And therefore the person doesn't need to be rebuilt by us" (*Timeline* 158).

Quantum physics. It gets worse.

The time traveler is not constituted in the other universe by our universe; the other universe is reconstructing it, reliably and perfectly.

> "It may be easier to understand," Gordon said, "by seeing it from the point of view of the other universe. That universe sees a person suddenly arrive. A person from another universe. . . . And that's what happened. The person has come from another universe. Just not ours."
>
> "Say again?"
>
> "The person didn't come from our universe. . . . They came from a universe that is almost identical to ours—identical in every respect—except that they know how to reconstitute [a time traveler] at the other end." . . .
>
> "She's a Kate from another universe?"
>
> "Yes."
>
> "So she's almost Kate? Sort of Kate? Semi-Kate?"
>
> "No, she's Kate [Gordon says]. As far as we have been able to tell with our testing, she is absolutely identical to our Kate. Because our universe and their universe are almost identical."
>
> "But she's still not the Kate who left here."
>
> "How could she be? She's been destroyed, and reconstructed" (*Timeline* 179–180).

How, indeed.

And do they transmit a soul? Is the person in the other universe just a facsimile of the original? Crichton says no, the person on the other end is the same person, just reconstructed in a different time and place. What about her memories? Her self-awareness? How does the MRI capture the biochemical reactions of thought and memory? Crichton doesn't say.

And what if the person is sent to a universe that can't reconstruct the data? That creates another hole in the plot. Niven wrote that the laws of probability preclude traveling to alternative universes. Say you throw a die. The odds of any one number coming up are one in six. Throw a pair of dice and the odds are one in thirty-six. If each probability represented a separate universe, then the chances of landing in any one of those universes would be one in thirty-six. How did the folks at ITC manage to just get that universe when they needed it since the number of possible universes would be almost uncountable?

Crucial to the plot is that the technology apparently doesn't work perfectly every time. The physicist who is found in the desert is one failure. He did not reconstruct perfectly on this end; again think of the teleport device in *Star Trek* which occasionally screwed up reconstructions.

Rules of the World

In every fictional world, there have to be rules. Gene Roddenberry, the creator of *Star Trek*, had strict rules about what could and could not be done in his universe, for instance the Prime Directive, which declared that the space travelers cannot interfere with a society or intervene in any way. They go only as observers. Many of the plots of that television program centered around instances in which the Prime Directive was morally challenged, or was at least seriously inconvenient. In time travel fiction, the rules are even more important because the plots are restricted by those rules and the rules themselves drive the plot. In *Timeline*, Crichton also creates rules. Sometimes they are mundane, but crucial to his story.

One rule is to not leave the vicinity of the machine and stay back in time more than an hour. Johnston gets in trouble because he broke both rules.

Another rule is to never bring any modern object into the past. No anachronisms. Crichton never explains why but two possibilities seem obvious. One, you don't want anyone in the modern world finding a modern object at an ancient site—modern bifocals or notes, for instance. (The company is trying to keep its technology secret). Having materials and objects that self-destruct or organically decay after some time would easily finesse that rule. ITC makes a few exceptions for defense and first aid, but never thinks to use organic materials that will recycle themselves. Two, the presence of modern technology could alter the timeline, although here Crichton seems inconsistent. If the presence of, say, a hand grenade (there is one in the plot) would alter history, why wouldn't killing a dozen French soldiers have even a greater effect? Or the introduction of Greek fire in a battle that would not have had it?

One rule he hinted at: never go back to an event in history that is important to a religion or a nation, lest you find out what you believe was only a myth. Do we really want to know that Abraham Lincoln had a squeaky voice and a head cold when he delivered the Gettysburg Address? What if time travelers discovered there never was an exodus of Hebrews from Egypt and that Moses was just a minor Egyptian prince of Hebrew ancestry, nothing more? What if Jesus turned out to be just another insignificant preacher in a city full of preachers whose reputation was immensely enhanced by great public relations and that there was no resurrection? Or that Mohammed was a weak-kneed puppet of his brother-in-law? Some of us probably wouldn't want to know any of that. By having them go back to what would have been a non-crucial battle in a war that lasted 100 years, Crichton avoids that inconvenience.

But all of this is fiction. One suspects Crichton had good fun putting it together and winding his way from physics to logical paradoxes to great leaps of scientific extrapolation. For at least some of the book, so does the reader.

JOEL N. SHURKIN is currently Snedden Chair in Journalism at the University of Alaska Fairbanks. He is former science writer at the *Philadelphia Inquirer* and at Stanford University, was founder of Stanford's science journalism internship program, and a freelance writer. He has written nine published books. He was a member of the team that won a Pulitzer Prize for covering Three Mile Island for the *Inquirer*. He is based in Baltimore.

ARTIFICIAL LIFE IN MICHAEL CRICHTON'S *PREY*

Larry Yaeger

In 1959 Nobel Laureate Richard Feynman gave a speech at the American Physical Society meeting entitled "There's Plenty of Room at the Bottom." Researchers in the field credit this talk as heralding the beginning of nanotechnology. This is a field that has been growing exponentially, not unlike the deadly nanoparticle swarms in Michael Crichton's Prey, *which Larry Yaeger discusses in detail.*

IT'S THE PROCESS, not the substrate.

That's essentially the claim, and might as well be the slogan, of so-called "strong" Artificial Life (ALife). The idea is simple: Life is best defined, both functionally and logically, by the processes that give rise to it, not by the materials in which those processes are implemented or embedded. The consequences of this insight are profound: Life—*real* life, be it naturally occurring or man-made—can be found or implemented in any system capable of sustaining those processes. ("Weak" ALife just says our computer algorithms can learn a lot from biology and would prefer to leave it at that.) Though many others, including Charles Darwin, Erwin Schödinger, John Von Neumann, Alan Turing, and John Holland, undoubtedly laid the groundwork, Chris Langton is the "father" of Artificial Life. He gave the field its name and made the clear, bold statement:

> The *big* claim is that a properly organized set of artificial primitives carrying out the same functional roles as the biomolecules in natural living systems will support a process that will be "alive" in the same way that natural organisms are alive. Artificial Life will therefore be *genuine* life—it will simply be made of different stuff than the life that has evolved here on Earth.

This is the core and perhaps most significant premise Michael Crichton posits to develop the deadly adversary in his novel *Prey*: evolving, self-reproducing swarms of nanoparticles that get cleverer and more dangerous—more *alive*—with each generation. Though not all scientists would agree, as far as I'm concerned we can cut to the chase on this one.... Life is indeed best understood as a complex network of processes; it is best quantified and measured in terms of information (the formal kind, due to Claude Shannon); and there is little doubt that *artifactual* (man-made) life is possible. Think about it this way: We already have an enormous number of examples of living and even intelligent machines, from bacteria to plants to insects to mammals to you and me. Of course living intelligent machines are possible; they're all around us! It's not much of a stretch at all to imagine a few more living intelligent machines that happen to be made of slightly different materials and that—oh, by the way—we happen to have had a hand in building or evolving.

Sure, naturally occurring, biological life on planet Earth has so far been composed almost exclusively of hydrocarbon chains floating around in liquid water. But as we will see, evolution—variation and selection—is *extremely* good at taking advantage of whatever resources are available. Our planet happens to have an abundance of carbon, hydrogen, and water, and these happen to be particularly suitable materials for sustaining living processes. Carbon provides strong atomic bonds in a great variety of forms and is the fourth most common element in the universe. Hydrogen is *the* most common element in the universe and oxygen comes in third. (Helium is the second most common.) Water, consisting of hydrogen and oxygen, then, is understandably plentiful, and liquid water, such as we find at the temperatures and atmospheric pressures found here on Earth, is a particularly convenient medium for holding and mixing the other materials of life. So evolution, as frugal as it is inventive, took persistent advantage of these common building blocks to coax generation after generation, species after species of creatures from the most convenient substrate here on Earth.

In fact, these basic elements are so common and they combine and recombine so readily, we probably shouldn't be surprised that organic compounds (most chemical compounds that include carbon, and comprise amino acids, proteins, nucleobases, and DNA), besides forming spontaneously in the lab, have now been found in meteorites originating here in our own solar system, in clouds of dust circling young stars some 375 light years away in the constellation Ophiuchus, and scattered throughout space and time, going back at least ten billion years—three-quarters of the age of the universe. It is entirely possible that hydrocarbon-based life inhabits a large proportion of our universe, and as a result of similar constraints producing similar results (something known as "convergent evolution"), alien life might even look a lot like terrestrial life. Though he doesn't call it that, Crichton invokes exactly this principle of convergent evolution to discuss the emergence and likely re-emergence of viral phages that are capable of invading the E. Coli bacteria used in the manufacturing process of the nanoparticles that comprise *Prey*'s deadly swarms.

It is also possible, however, that alien life might be based on substantially different chemical substrates. It has often been suggested that, under the right conditions, silicon might replace carbon, going back to scientific speculation by German astrophysicist Julius Scheiner in 1891 and fiction by H. G. Wells in 1894. Similarly, liquid methane, nitrogen, carbon dioxide, or even sulfuric acid might substitute for water. In fact, to avoid silicon's strong tendency to be *sequestrated* (soaked up) into silicon (di)oxide, silicon-based life might have to evolve in an environment largely lacking in oxygen and therefore lacking in water. To find those other liquids, and for silicon's weaker (than carbon) bonds to provide the backbone of life, the environment would probably also have to be quite cold compared to anything on Earth. Saturn's moon, Titan, for example, might very well provide an environment suitable to silicon life.

But, if one accepts the premise that life really is a process, not the specific materials in which that process is embedded, then there is another, radically different substrate that might support life right here on Earth: digital computers. Crichton posits the possibility of digital, artificial life, of a particularly unique and interesting kind, in his nanotech swarms. The study of such lifelike and biologically inspired processes in computers is the domain of Artificial Life, a relatively new scientific discipline. It was officially born under that name in 1987, at the first Artificial Life conference, convened at Los Alamos National Laboratory by computer scientist Chris Langton, but its roots extend back to at least the time of Darwin and many of its themes were broached by John Von Neumann and Alan Turing. In fact, the very first electronic computers in existence were immediately turned to studies of "symbiogenesis" by Nils Barricelli in 1953 on the IAS computer at Von Neumann's Institute for Advanced Study, and population genetics by Alex Fraser in 1956 on the SILIAC machine at the University of Sydney, Australia.

There are probably enlightened individuals who understood and embraced the mechanistic and fundamentally deterministic nature of living systems even earlier, but late Victorian iconoclast and author of *Erewhon*, Samuel Butler, wrote in 1880's *Unconscious Memory*:

> I first asked myself whether life might not, after all, resolve itself into the complexity of arrangement of an inconceivably intricate mechanism.... If, then, men were not really alive after all, but were only machines of so complicated a make that it was less trouble to us to cut the difficulty and say that that kind of mechanism was 'being alive,' why should not machines ultimately become as complicated as we are, or at any rate complicated enough to be called living, and to be indeed as living as it was in the nature of anything at all to be?

Here, Butler identifies a particularly important issue in understanding the nature of living systems, which is that there really is no good definition of what it means to be alive and that "being alive" may, in fact, be nothing more than a convenient shorthand for describing systems above a certain threshold of complexity. (Curiously, Butler was also a frequent attacker of Darwin and Natural Selection in favor of Lamarckian "transmission of acquired characters"—the idea that offspring acquire traits most frequently used by their parents. This idea of "use inheritance" had held sway until Darwin, but has never actually been found in natural biology. Lamarckian inheritance was finally demonstrated to be nearly impossible with the identification of DNA as the primary mechanism of inheritance, since there is no way for learned or heavily used behaviors to be reverse-transcribed back into the genome. Butler also came to prefer thinking of this continuity of inorganic parts and living systems as imbuing some essence of life on those inorganic parts, rather than implying any lack of life for organic systems.)

Imminent British biologist C. H. Waddington, in decrying "vitalism" (the notion that there must be something non-physical animating all living things) in 1961's *The Nature of Life*, wrote:

> Vitalism amounted to the assertion that living things do not behave as though they were nothing but mechanisms constructed of mere material components; but this presupposes that one knows what mere material components are and what kind of mechanisms they can be built into.

Here, Waddington makes a vital and elegant point: that our knowledge of the full range of possible behaviors of material mechanisms is woefully limited and will almost certainly be underestimated unless we continue to study and understand those possibilities.

The essence of ALife, then, is the study and particularly the *synthesis* of complex, emergent, pseudo-biological mechanisms. It is a broad interdisciplinary field, encompassing agent-based modeling, evolutionary algorithms, neural networks, multiscale cellular modeling, cellular automata, L-systems, swarming, flocking, and more. Related work has produced methods for routing Internet traffic, improving automobile traffic flow, managing people flow and congestion in crowd-panic situations, searching for documents on the Internet, optimizing airfoil shapes, performing general purpose function optimization, and much more. In the area of robotics, ALife research has produced flying robots, soccer-playing robots, and robots that can work around damage to their own bodies. As a kind of "theoretical biology," ALife research allows the study of co-evolution, parasitism, cooperation, kin selection, and other critical features of evolutionary biology and behavioral ecology. ALife has drawn on the study of biological immune systems to develop new forms of computer security and has taken lessons from artificial immune systems back into the realm of biology, suggesting novel, testable hypotheses about the functioning of natural immune systems. ALife models can even yield insights into the origin of biological life and are beginning to explore the possibility of evolving new forms of life from scratch. (Kim and Cho have an extensive review of ALife applications in the January 2006 issue of the *Artificial Life* journal, if you're interested.) As Langton put it in his introduction to the proceedings of the first Artificial Life conference:

> By extending the empirical foundation upon which biology is based *beyond* the carbon-chain life that has evolved on Earth, Artificial Life can contribute to theoretical biology by locating *life-as-we-know-it* within the larger picture of *life-as-it-could-be.*

In writing *Prey*, Crichton drew heavily on key insights from the field of ALife. In particular, his intelligent and predatory swarms are based on a number of central premises, almost all of which inform and are informed by ALife research:

1. Biological / genetic engineering
2. Nanotechnology
3. Self-organization
4. Emergence
5. Swarming or flocking
6. Predator-prey behaviors
7. Evolutionary algorithms

It probably should be said that Crichton unfortunately gets a fair number of the scientific details wrong in all these areas, and he has rightly been taken to task for it elsewhere. I will draw attention to some of the more glaring mistakes, but for the most part I intend to focus on the very real science behind the book's ideas. Please do not take this as an endorsement of the reality of the book's scenario. While I think every one of the areas of science being drawn from here is fascinating and important, I don't think the pieces fit together or work in quite the fashion Crichton imagines in his novel. But I don't feel a need to attack the man for his scary portrayal of corporate greed and scientific misconduct gone horribly awry.... He was writing a thriller! There had to be conflict, just like there had to be a beginning, a middle, and an end. Okay, he stretched (some would say broke) the truth, and perhaps tried to intimate a greater degree of scientific authenticity than the book deserves with his little science lessons throughout and the reference section at the end. But Crichton did do his homework on this; that reference section is pretty impressive (if I do say so myself, what with one of my papers listed). I suspect those details were part of a calculated effort to make the book *feel* realistic, whether it is entirely or not. And by and large it worked. The book scared people, as intended.

The fundamental construction of *Prey*'s predator swarms is based on bioengineering and nanotechnology, embedded in a manufacturing process based approximately on modern computer chip fabrication facilities. Genetic modifications to E. Coli bacteria cause them to produce key component molecules which then self-organize into nanotech *assemblers*, with the assemblers constructing the final nanoparticles that comprise the swarm. Those assemblers, in turn, have been designed in such a way that they adhere to the bacteria's cell membranes, ostensibly so as to prevent the "heavy" assemblers from being drawn away from the lighter bacteria and constituent molecules and thereby reducing the efficiency of their manufacture. There are details here that one can definitely quibble with, but by and large the ideas are sound.

The use of gene-tailored bacteria in the manufacturing process is more than reasonable and is a technology that, though still in its infancy, is growing by leaps and bounds. As a means of producing complex molecules, it may ultimately be without parallel (until general purpose nanotech assemblers become a reality, if they do). As early as 1978, genetic engineering was used to modify E. Coli bacteria to produce human insulin, and the resulting insulin was awarded FDA approval in 1982. Genetic modification of hamsters has been used to make them produce the human hormone regulating red blood cell production, *erythropoietin*. Genetically modified foods are a controversial topic, but offer such possibilities as the reduced application of herbicides, increased crop yields in the face of global warming, elimination of whole classes of malnutrition, and the delivery of oral vaccines through the consumption of fruits and vegetables.

Of course, it should not escape one's attention that the use of bacteria in the nanoparticle manufacturing process provides a nice, sound reason within the context of the story, for the swarm to require biomass in order to reproduce. This lends the story one of its eeriest, most disturbing aspects: these swarms literally prey on animals, including humans, as a natural part of their life cycle. This is particularly true of the "wild" type that immediately kills and consumes their prey, but is also the source of the emaciation and dissipation of the human hosts of the "benign" type.

Since the bacteria were already identified as crucial to the nanoparticle production pipeline, it is curious that Crichton chose to introduce and call attention to the whole idea of "reapplying" the bacteria—making the nanotech assemblers adhere to the bacterial cell walls to deal with weight differences. I imagine he felt this strengthened the necessity of including the bacteria in the swarm, since one might reasonably argue that once you have functioning assemblers you could continue to create additional swarms without the bacteria. However, with a fixed number of assemblers, the reproduction rate of the swarm would have been fixed, and he could have as easily posited an awareness of this on the part of the (seriously misguided) scientists who released the swarms' raw material into the environment, thus an inclusion of the bacteria in that release, so as to allow the assemblers themselves to reproduce. This would have been sufficient to allow him to justify the inclusion of the bacteria in the swarms' composition.

As important as these flesh-eating bacteria are to the horror of the story, and even though I think you can make a good case for their inclusion in the self-reproducing swarms, they are also part of the biggest logical hole in the story. The grand, dramatic conclusion depends on these bacteria being destroyed by a phage (a virus that invades bacteria). But the nanoparticles, as they were originally designed at least, don't even need the assemblers to function, much less the bacteria. So, in theory, the benign swarms infesting humans could have lost their bacteria and kept right on functioning. They just wouldn't have been able to reproduce until they shed their dead, infected bacteria and found some new ones. Crichton could have arm-waved a bit and claimed that the swarms had evolved a more intimate relationship with the bacteria, or that nanoparticles were constantly expiring and being sloughed off like dead cells, hence they needed constant replenishing, but he didn't.

Much earlier in the story, since the original function of the nanoparticle cameras didn't require assemblers and bacteria and since they're the only element of the swarm that is being driven to actually swarm, it's hard to imagine that simply dumping a bunch of the materials out of doors would keep the nanoparticles, the assemblers, and the bacteria in close enough and continuous enough proximity to ever allow the first swarm to reproduce. As powerful as evolution is, species go extinct all the time. Conditions have to be right for life to persist and the conditions for it to *begin* are even more stringent. It happens, of course. We're living proof. But this is one of those places you have to just ignore the facts and do the suspension-of-disbelief thing in order to get on with the story.

The posited scales and weight differences are another place where the science in *Prey* breaks down. It seems unlikely that even a large number of molecules brought together to make an assembler, no matter how complex or what their composition, would weigh more than an entire bacterium composed of a trillion (or so) molecules. So, the whole assembler > bacterium > molecule weight ordering seems out of whack (and, as previously stated, unnecessary to keep bacteria in the swarm reproductive mix). But, then, the nitty-gritty scientific details get a bit murky at this scale anyway. The main character, Jack Forman, ruminates about "a typical manufactured molecule [consisting] of 10^{25} parts" (*Prey* 128). Granted, something consisting of 10^{25} *anythings* probably would be heavier than an E. Coli bacterium. However, if he really wants to talk about the composition of molecules, "parts" would have to mean atoms, and nothing that consists of 10^{25} atoms would normally be considered a molecule, except possibly in the atypical case of certain crystalline structures such as diamond or graphite sheets in which all of the atoms are the same (Carbon). Molecules usually consist of a few atoms to a few tens of thousands of atoms. Even the largest biological molecule, the monster DNA, consists of roughly 10^{11} atoms (for humans, that is; *Psilotum nudum*, the "whisk fern," currently has the largest known DNA molecule, weighing in at about 100 times the size of human DNA). Perhaps Crichton meant to refer to a "molecular machine," rather than a "molecule," though still, 10^{25} atoms is a bit to the large side, being something like ten times the number of atoms in the smallest known insect, a parasitic wasp known as *Dicopomorpha echmepterygis*. (Admittedly, this is a *very* tiny wasp, being smaller than a large single-celled paramecium, and that estimate of the number of atoms in the wasp is *extremely* rough.) However, neither a lower estimate for the molecular machine's size, comparable to the number of atoms in a single modest-sized cell (roughly 10^{17}, say), nor Crichton's 10^{25} estimate of atoms remotely fits the description of the camera as being "one ten-billionth of an inch in length." That would be 10^{-10} inch, which works out to 0.0254 Å—much smaller than a single atom! So I'm afraid we're just going to have to agree to not sweat the small stuff.

Speaking of small stuff, nanotechnology is another burgeoning, if nascent, field. Dealing with materials and devices on the scale of one to 100 nanometers (ten to 1,000 times the size of a single atom), nanotechnology embraces a wide array of subjects from materials science to drug delivery to novel computer displays to food packaging to quantum computing. Basically, matter behaves very differently at the atomic scale than it does at the everyday macroscopic levels we are used to, and there are both efficiencies and genuinely unique properties available at these small scales that we are just beginning to learn how to exploit. Imagined by Richard Feynman and popularized by Eric Drexler, nanotechnology is really just beginning to yield results. Yet, as of 2007, it has been estimated that there are already some 360 consumer products incorporating nanotechnology in some form. A number almost double the estimate for the previous year. While flesh-eating swarms may not be a particularly common concern, nanotech researchers have long worried about the infamous "gray goo".... This being the idea that self-reproducing assemblers might turn all matter into copies of themselves, producing a homogenous sludge where once lived plants, animals, and humans. However, making such a self-reproducing assembler is remarkably difficult and is something no one would make on purpose, so the gray goo scenario is probably not all that serious a threat. Much more worrisome is what I've been thinking of as "gray lung".... That is, nanotechnology, by its very nature, creates materials that are at a scale small enough to be highly biologically active. Asbestos and soot are such health problems in part because of their small size, getting as tiny as about 10 nm in diameter, yet they dwarf fullerenes, buckyballs, and carbon nanotubes, the most common nanotechnology products, whose diameters are only about 1 nm. Never mind being able to reach deep lung tissue more effectively, these materials are of such a scale that they can interact chemically with whatever tissue they end up in. There is already empirical evidence that fullerenes in water can produce brain and liver damage in fish and is otherwise toxic to water fleas and fish. And computer simulations suggest that buckyballs can bind to and deform DNA, potentially causing untold genetic damage. Even if there is *never* an industrial accident involving nanotech particles (can you say *Exxon Valdez*?), as nanotechnological products reach the end of their life cycles and are disposed of, we may not be able to control their dispersal into the environment. Nanotoxicity is a matter of real concern to scientists today and we can only hope that governments, corporations, and scientists behave responsibly and treat this issue with the respect it most definitely is due.

Though he doesn't refer to it by that name at the time, Crichton also calls on the principle of self-organization to help explain the success of the nanoparticle production process outside of the clean room factory. Jack's former friend and swarm infectee, Ricky Morse, comments, "The component molecules go together quite easily" (*Prey* 137). These component molecules combine to form the final nanotech assembler. It is a valuable contribution to the production process for the molecules to self-organize and self-assemble. It is, of course, also a valuable contribution to the storyline for the nanoparticle production process to work outside the lab!

Self-organizing systems are like chemicals that react without the need to do anything except bring the chemicals, the individual parts, into close proximity. Despite the inviolability of the second law of thermodynamics that says disorder, entropy, and simplicity always and only increase on a global scale, self-organizing systems can take advantage of external energy sources or internal energy flows to *locally* generate additional complexity and create islands of reduced entropy. In a very formal sense, Claude Shannon showed us that negative entropy *is* information. So, self-organizing systems create information from energy and sustain that information in the face of the larger system's inevitable path toward the cold, dark, thermodynamic death of the universe. Even before Shannon had defined the relationship between entropy and information back in 1948, and well before physicist Edwin T. Jaynes had elucidated the intimate connection between Shannon entropy and physical entropy in 1957, Erwin Schrödinger, in his influential 1944 manuscript, "What Is Life?," put forward the idea that life itself is based on and best characterized as islands of "negative entropy," i.e., it is the specific and unique nature of living things as self-organizing systems to *locally* defeat the second law of thermodynamics, creating order and complexity from chaos. (In that publication,

Schrödinger also put forward the idea of a complex molecule providing the genetic code for living organisms, which inspired Watson and Crick to research the gene and led to the discovery of the structure of DNA.) I believe Schrödinger's observation is a true and profound one, and suggests the way forward in understanding life and perhaps even quantifying it. Information theory, à la Shannon, provides the tools for identifying what is unique to living systems and assessing their behaviors. I have been exploring the application of an information theoretic measure of neural complexity to analyze the evolution of complexity in neural network-based artificial life with Olaf Sporns at Indiana University. Chris Adami uses information theory to quantify the evolution of complexity in simulated bacterial colonies and to identify structurally and functionally related nucleobases in biological DNA. John Avery's excellent book, *Information Theory and Evolution*, expands upon Schrödinger's insight in a clear, concise fashion. A broad array of scientists is embracing this simple idea that life *is* information. And information is accreted and sustained through self-organization.

Crichton does use the term "self-organization," even getting around to abbreviating it as "SO" when he discusses the swarming or flocking behavior of his nanoparticles. Swarming or flocking is indeed a kind of self-organizing behavior. In fact, it is remarkably easy to obtain flocking behavior, given how complex it appears to be when observed in action. Craig Reynolds, the inventor of flocking algorithms, sat and watched birds lifting off from a grassy field, flying in coordinated groups for a while, then returning en masse to the ground. He wondered what controlled this beautiful, seemingly complex and coordinated behavior. Does it require a lead bird? How smart do birds have to be to behave this way? The answers turned out to be no and not very. His elegant solution consists of three simple rules:

1. fly toward the average position of the birds nearby you;
2. orient yourself to match the average orientation of the birds nearby you;
3. avoid bumping into other birds.

That's it! Crichton explicitly referred to the first and third rules only, and, in fact, you can get chaotic, interpenetrating, colliding swarms using just those two rules, although it turns out that you also need the second rule to get the organized, graceful schooling and flocking seen in birds and fish. In 1986 Reynolds used a computer to model what he called "boids" (synthetic birds) and showed that they produced quite believable flocking behaviors given those three simple rules, publishing his work in 1987 at the SIGGRAPH conference (Special Interest Group in computer GRAPHics). He also combined flocking with obstacle avoidance and goal-seeking behavior and experimented with evolving sensor placement and motor control of artificial agents learning to steer through complex corridors. Reynolds-inspired flocking algorithms have since been used in scientific research, high-tech art projects, and a number of Hollywood movies including *Batman Returns*, *The Lion King*, and Crichton's own *Jurassic Park*.

Flocking and swarming are excellent examples of self-organization and of *emergent* behavior—a whole that seems to be more than the sum of its parts, even though it is entirely defined by those parts. (In fact, Chris Langton named his primary second-generation ALife research platform "Swarm.") *Emergence* is a key concept in ALife and in the study of complex systems, including biological life and human intelligence. Crichton uses emergence to explain (or deftly and reasonably avoid explaining) a number of key story elements, in particular a number of behaviors of the predatory swarms. He correctly uses it to discuss flocking and swarming. He also gets it quite right in discussing the distinction between traditional, symbolic, rule-based AI (both affectionately and pejoratively, depending upon your point of view, known as GOFAI, for Good Old Fashioned Artificial Intelligence) and parallel distributed processing (PDP) or ALife approaches to AI. As Jack ruminates, GOFAI is "top down"—all knowledge, all final, higher-level behaviors are specified by the programmer. PDP and ALife are fundamentally "bottom up"—low-level behaviors and models are used, with the resulting higher-level behaviors manifesting as emergent properties of the system. Sometimes those emergent properties can be predicted. Sometimes they can't—a fact Crichton depends on to explain some of the swarms' unexpected actions. But, contrary to what some sloppy thinkers might tell you, there is nothing mysterious or otherworldly about emergent behaviors, and emergent behaviors are not immune to scientific inquiry. In discussing *Intelligence as an Emergent Behavior*, Danny Hillis observes, "Anyone who looks to emergent systems as a way of defending human thought from the scrutiny of science is likely to be disappointed." In fact, one of the great strengths of the bottom-up approach is that you can apply what limited knowledge you might have about simple, low-level systems to help you develop a better understanding of higher-order, complex systems. The everyday properties we ascribe to gases and liquids, like temperature, pressure, and viscosity, are the emergent properties of the low-level interactions of molecules. There is reason to think that intelligence is similarly the emergent property of the low-level interactions of neurons. So, emergence actually helps to logically guide our understanding and modeling of complex systems by serving as a means of identifying and distinguishing between different levels of organization. And just as Hillis was able to evoke complex fluid dynamics such as laminar flows, vortices, and turbulence out of a remarkably simple model of molecules—unit mass particles on a hexagonal grid exhibiting "billiard ball" physics—so he (and others and I) speculates you might be able to evoke intelligence out of surprisingly simple models of neurons. This is an idea that, as Hillis wryly puts it, "has considerable appeal to the would-be builder of thinking machines." Similarly, the seemingly complex flocking behavior of birds, schooling behavior of fish, and swarming behavior of *Prey*'s deadly swarms turn out to be an easily obtained emergent property of surprisingly simple low-level behaviors.

It turns out that flocks, schools, and swarms offer real value to the individual members of the group. The complex, chaotic motion of swarms, especially as they are dispersing, can confuse predators. Computer simulations have demonstrated the evolution of schooling behavior in fish in response to predation, and have shown that predator efficiency is as negatively impacted as members of the school might hope. Plus, as Richard Dawkins would undoubtedly point out, even the unfortunate member of a school, flock, or herd that gets culled and eaten is probably closely related to at least some of the members of the group that got away, hence its genes will live on. Living in such a group also makes it relatively easy to find a mate, even as the group moves around to look for new and varied sources of food. For artificial systems, swarms can offer a convenient redundancy and a sort of strength in numbers. In simulations eerily reminiscent of *Prey*'s swarms, John Barker, who does electronics and nanotechnology research at the University of Glasgow, Scotland, has been modeling what he calls "smart dust"—swarms of thousands of minuscule wireless sensors that swirl and navigate across artificial landscapes by switching their shapes between smooth and rough, dimpled forms. Sound familiar? In 2007 Barker simulated the release of 30,000 such particles on a Martian landscape and found that a simple set of rules embedded in each particle was sufficient to allow 70 percent of the motes to navigate a 20 km course. Real-world smart specks don't qualify as nanotech yet, ranging from existing commercial smart sensors that are 25 mm in diameter—approximately the size of a quarter—down to experimental devices that are only a few square millimeters in size, but nanotech versions of these smart-speck swarms may one day explore strange new worlds and boldly go where no man has gone before.

However, Crichton goes more than a little astray at this point in his story and in his science, imbuing self-organized behavior with properties that are really only associated with directed intelligence. These are not the same thing, even though Crichton uses them as if they are. To his credit, he does arm wave a bit about certain intrinsic "if the current behavior isn't working, try something else" behaviors having been built into the PREDPREY program that is being executed by each of the nanoparticles. And ostensibly PREDPREY provides goals for the swarm. But he is apparently a bit confused about all this, making directly conflicting statements and observations about PREDPREY versus the swarm's behavior. He states a number of times that the primary purpose of PREDPREY was to make a swarm stick to an existing goal, yet he says it might not do that without "reinforcers," which apparently aren't built into PREDPREY. Okay, maybe it's not very good at its intended job, but even performing poorly at sticking to a goal isn't anything like the adoption of entirely new goals, much less intelligently selecting sub-goals to accomplish an overarching goal, such as when the swarm first blocked the door so that Jack could not return to the lab (the intelligence of which Jack remarked upon). Crichton also had Jack muse, "I wanted to know why [the swarm] had pursued the rabbit. Because PREDPREY didn't program units to become literal predators" (*Prey* 155). Yet later Jack observed, "I knew the swarms were programmed to pursue moving targets if they seemed to be fleeing from them" (*Prey* 165). So, according to its inventor, PREDPREY both does and doesn't chase fleeing prey. Ah.

Part and parcel of this confusion is a claim that the self-organization aspect of these swarms somehow produces genuinely intelligent behavior that rapidly adapts to changes in the environment, adopts new goals, and calculates intermediate goals to attain long-term goals. A bit of randomness and exploration is indeed a key part of certain learning algorithms, and self-organization can produce some startlingly complex emergent behaviors. But as with Crichton's assemblers, which tend to assemble themselves regardless of environmental conditions, self-organization tends to foster adaptability only in the sense that the same results are achieved under varied circumstances, and not in the sense of innovating entirely new results. Indeed, self-organizing behaviors are frequently stereotypical and relatively fixed. They may be adaptive within the modest range of environmental variability they were exposed to when they were evolved, but they are generally unable to adapt to radically different environments or challenges. Birds and fish exhibit self-organized swarms, but they don't build birdhouses or aquaria. Ants and bees, for all their complexity and success at mastering certain environments, cannot adapt to other environments, nor are they likely to ever organize and send an envoy to the U.N. Self-organization is not synonymous with or a guarantee of intelligent behavior.

Okay, maybe those first objections to PREDPREY's described nature are really just quibbles about minor authorial gaffes, not bad science, per se. And while self-organization doesn't guarantee intelligence, it certainly might contribute to it. However, there seems to be a fundamental misunderstanding of what predator-prey interactions really consist of. Basically, predator-prey interactions are what you probably think they are: one species chases and kills another, while the other species tries to get away. That's pretty much it, though there are some interesting dynamics that come out of such interactions. For example, if the predators are too efficient or there are too many of them, then they can wipe out all the prey, at which point the predators are out of luck as well, because they've killed all their food and will subsequently die of starvation. Predators and prey can also get into population boom and bust cycles, called *Lotka-Volterra* cycles, named after the two scientists who independently developed equations to describe these dynamics back in the mid-1920s. What happens is that as the prey become more plentiful, the predators can more easily find food, so the predator population begins to grow as well. But once the predator population gets large, they eat so many prey animals that the prey population begins to dwindle. As the prey population drops, the predators go hungry, so their population also begins to decline. But a declining predator population allows the prey population to rebound. Which allows the predator population to rebound. And so on (and on, and on). What, you might ask, does this have to do with setting or maintaining goals for predators, prey, or their combination? (Good question!) Answer: nothing. Now, actually, if you combine predator-prey interactions with evolution in just the right way so that predators act as natural selection's "reapers" by either removing misbehaving particles from the swarm or, more likely, reprogramming them, then it is at least conceivable that this combination might help a swarm maintain its goals (more on this in a bit), but even in this situation it is not clear how the swarm would obtain new, more complex goals or develop anything resembling intelligence.

Of course, under certain circumstances and with enough time, evolution certainly can select for changing, increasingly complex behaviors and higher intelligence—it has done so once, or you wouldn't be reading this—but when Jack is crediting all these adapting behaviors to the swarms, he doesn't even know they've been reprogrammed to evolve, according to the story. Any evolution under those circumstances would have to be due to errors in the reproductive process (variation) and subsequent selection. And evolution, being completely unguided, requires there to be an evolutionary advantage to a new variation in order for it to be selected. Given enough time, swarms could conceivably evolve to better maintain their forms and even to increase their intelligence. But there is no survival or reproductive advantage to some of the instantaneously "evolved" behaviors that Jack credits them with, such as zig-zagging search patterns and their seek-and-destroy behavior when the human characters are hiding in the storage shed or the cars. Perhaps it is more accurate to say that there *is* an evolutionary advantage to these behaviors, but not on the time scale to which the story credits their adaptation. If a zig-zagging behavior helps a swarm find food, then swarms that zig-zag will survive and reproduce. But there is nothing about zig-zagging that improves the fitness of the swarm or its nanobots *before* it has found the food. Crichton alternately credits self-organization and evolution with the ever-improving behavior of his swarms, but neither really works quite as described. Evolution is closer to the mark, but it still needs a selection pressure that simply isn't present under the circumstances.

That said, I can *almost* come up with a couple of different rationalizations for the evolving, adapting behaviors Crichton describes.... Suppose the predator part of PREDPREY was able to assign fitness, in an evolutionary sense, to particle behaviors, and it was able to do this in each new situation that the swarm encountered. Then suppose the predator particles attacked only those prey particles that had a low fitness—that failed to contribute to the attainment of PREDPREY's goals. Presumably the predators themselves would be *co-evolving*, their fitness being determined by their success at winnowing the herd of prey particles. You'd then have a rapid evolutionary process, now operating on machine timescales instead of natural ones, which explicitly supported the goals of PREDPREY. One problem with this scenario is that there is no mechanism to evolve PREDPREY's goals, so, at best, you'd have a system that very efficiently imposed a strictly fixed set of goals on the swarm, and none of the worst outcomes of the novel's speculation could ever have come to pass. Perhaps a more damning problem with this scenario is that it requires PREDPREY itself to be intelligent and nearly omniscient, knowing the behavior of every particle and assigning it an individual fitness. Even if the knowledge of particle state and the assignment of fitness could be distributed to each of the individual particles, no goal that required anything resembling learning or directed intelligence would be possible. Still, with a very limited, precisely specifiable goal such as "limit the swarm to a certain maximum diameter," this is how I imagined PREDPREY might work until Crichton indicated evolution was not a part of the original PREDPREY design.

The other idea I came up with for explaining a swarm's intelligence would require it to have developed the equivalent of a neural network, where each particle acted like an individual neuron, and to have begun using the network like a nervous system—like a brain—to intelligently control its behavior. This isn't likely without a suitable low-level model of neural behavior embedded in the particles and adequate communication between particles. And how the network organization (the wiring diagram of the artificial brain) is defined and then maintained as the particles move about so frenetically is, *ahem*, an open problem. But at least one could then begin to speculate about directed intelligence and real-time learning. However, nothing like this was even vaguely suggested by Crichton, so never mind.

Crichton is absolutely right, however, about the extreme power of evolutionary algorithms. Not that we can't tell that simply by looking around us at all the fruits of such a process. But scientific research into the functioning and the application of evolutionary algorithms has shown them to be remarkably effective and robust. For optimization problems where you're looking for the best values of a set of parameters, a Genetic Algorithm (GA) can hardly be beat. Oh, if a problem is simple and well-behaved enough, then either analytical solutions or so-called "hill-climbing" solutions might do well and even beat a GA, at least in computational efficiency. But for most real-world problems, an analytical solution simply isn't possible. (Try writing down the equations for the optimal shape of an airfoil!) And usually the "error surface," that becomes a "fitness landscape" when inverted and defines just how each possible set of parameter values performs for the problem, is fraught with troubles for hill climbers. Multiple peaks and valleys in such a landscape mean that hill climbers get stuck on "local optima." That is, they happily climb a small hill and, once at the top, discover that no matter which direction they turn, things only get worse, so they get stuck on this small hill, even when there's a massive mountain peak just a short valley away. Hill climbers also need to be able to compute *derivatives* (the slope of the landscape) everywhere in order to figure out which direction to go next. GAs use entire populations of individuals which are capable of exploring many different peaks in parallel, thus finding the highest peak instead of getting stuck on some small hill. Also, GAs are not dependent on derivatives of the error surface. All they need is the individual's immediate fitness—the height of the landscape at that particular point. So, in many ways GAs are much easier to implement and more robust than most other optimization techniques.

Another neat aspect of GAs (and Genetic Programming and most evolutionary algorithms), perhaps contrary to widespread opinion, is that they don't usually rely all that heavily on mutation (random changes in the genes) to achieve their goals. Instead they take advantage of something else that nature discovered called "crossover." What crossover means in GAs is that when two "parent" individuals produce an offspring, you take some genetic material from one parent, then you "cross over" and start taking genetic material from the other parent for a while. You might cross over between parents a number of times this way. In the end you have one complete set of genes that consists of a combination of gene sequences from both parents. (Genetic crossover in humans and other biological entities is a little bit different in detail, but similar in spirit.) The reason this is important is that it lets evolution try out different combinations of fundamental building blocks. You might pick up one good segment of genes from one parent and a different segment of good genes from another parent. It's mix and match. Of course, you might pick up a bad combination of genes, but in that case, the resulting offspring wouldn't be very fit and wouldn't produce many offspring of its own, so the bad combination will die out, while the good combination lives on. That's the whole idea behind natural selection. In fact, John Holland showed that this effect of mixing and matching what he called "schemata" is so powerful that mutation is relegated to basically little more than preventing evolutionary dead ends, or what you might think of conceptually as stirring the pot. It's crossover that does the heavy lifting, putting pieces together in different combinations for natural selection to act on until it finds the right combination.

Because of the population effects, the mix and match of crossover, and the low-level variation introduced by mutation, GAs are *very* good at exploring even the most complex of fitness landscapes. Evolution is astonishingly good at finding solutions to problems, including solutions that a human might never have thought of, and might not even want! I refer to this as evolution finding "any little niche" in a class I teach on the application of ALife to the problem of AI. Anyone who has created their own ALife program has stories about how evolution found some bug in their program, or some unintended and unwanted behavior that was sufficient to solve the problem the agents of that world were faced with. To make matters worse, once evolution finds a solution that works, no matter how simple and no matter how much you wish it would do something else, unless there is some pressure to change, it will just keep doing this unremarkable or unwanted thing. The problem is that, like the difficulty faced by hill climbers looking for the highest point on a complicated landscape, any small change from this simple-but-adequate solution—this local optimum—is likely to be less fit than the original. And any large change is likely to be entirely maladaptive. As Philosopher of Science Daniel Dennett puts it, "The cheapest, least intensively designed system will be 'discovered' first by Mother Nature, and myopically selected."

Two of my favorite any-little-niche stories come from Karl Sims, author of an inventive world of "blocky creatures" which he first evolved to walk (as well as crawl, hop, slither, and swim) and later evolved to compete over a block placed between two creatures. When he was trying to get the creatures to simply move across a landscape, he made the fitness function of his evolutionary algorithm be simply the distance that their center of mass moved horizontally along the ground between the beginning and end of a run. That should do it, right? But what he got instead of walkers or the like were faulty towers. That is, his program routinely evolved creatures that consisted of nothing but tall blocks piled atop each other that did nothing at all until right near the end of the run, at which point they simply fell over. The act of falling over moved their center of mass a significant distance horizontally (as well as vertically, of course), without all that nasty messing about with limb motion, walking, and the like. Of course, this was not what he had in mind or what he wanted. Simply falling over was no fun at all! Thankfully, a few simple constraints on vertical growth and, voila, he got walking, crawling, hopping, slithering, you name it. Neat stuff. Look for the videos on the Web.

But evolution wasn't done frustrating Karl yet. In order to let the creatures walk, he had to have a fairly decent "physics engine": software that defines how the world works, providing gravity, friction, momentum, and the like. In order for the simulated physics to work, his software had to know when one object was in contact with another object, so pushing against the ground in one direction could provide motion in the opposite direction, for example. Figuring out when one computer object is touching another is called "collision detection." and Karl built a simple collision detection algorithm into his system for this purpose. Unfortunately it had a bug: It turned out that when two objects initially collided, one would start to go right through the other. That part was actually expected, and the collision detection algorithm, upon discovering this situation, would identify this as a contact point between the two objects and then nudge the objects apart just enough that they would no longer interpenetrate. But it turned out that if the two objects interpenetrated by a *very tiny* amount, Karl's collision detection algorithm would effectively try to divide by zero. As a result, instead of being nudged ever so slightly, one of the objects would be launched to infinity! Then some bounds-checking code would kick in and, instead of being placed at infinity, the object would be placed right at the edge of the world. So, when this bizarre error condition occurred, one of the objects would move the absolute maximum distance possible! But sheesh, this is such a uniquely specific and precise failure mode that surely evolution would never find it, right? (Remember, the Intelligent Design lot insist that a bacterium is too much for evolution to handle.) Well, if you thought that, then you (as well as the IDiots) would be wrong. Creatures in Karl's world evolved to rub their own limbs together, slowly, slowly, slowly bringing them ever closer together until BAM—their two limbs would just barely interpenetrate by the required miniscule amount, and suddenly they'd teleport to the edge of the world. And, annoyingly, being the fittest creatures in the world, once evolution discovered this bizarre behavior, that's pretty much all it would ever produce. Of course, Karl sorted it out, fixed the code, and evolved some of the most interesting creatures the field of ALife has seen.

Bottom line, as Crichton suggests and the fundamental premise of *Prey* depends upon, evolution will both A) surprise you and B) do amazing things, without any guidance or intervention of any kind. All it takes is an environment that lets things interact with each other and some low-level building blocks that are able to make other low-level building blocks. Probably the first biological reproducing systems were so-called *autocatalytic* sets of chemicals. It's not even necessary that some individual thing can make a copy of itself. It's enough that, say, chemical A can help *catalyze* (facilitate) the production of chemical B, while chemical B can help produce another chemical, which helps produce another chemical, which eventually gets around to producing more of chemical A. Voila! A closed loop. An autocatalytic set. Biophysicist and complex systems researcher Stuart Kauffman showed that any moderately sized set of interacting components is statistically essentially *guaranteed* to form such a closed loop—to produce an autocatalytic set. Then, whether it happened at a deep sea hydrothermal vent, the small channels of clay formations in shallow pools, inside oil drops, or just floating around in some primordial soup, it turns out that these sets of chemicals are more efficient if they aggregate (hang about in the same little area), because the chemicals are more statistically likely to interact—they can find each other. And, boy, if you do get these chemicals concentrated inside an oil drop or any kind of semi-permeable membrane, this little proto-cell turns out to be more efficient yet at reproducing its constituent chemicals. In fact, as the sack of chemicals gets too big, it pinches in two, and now there are two little proto-cells busily churning out more chemicals and proto-cells. Then the race is on, exponentially and then some.... If you bring a few cells together, they can specialize and make a more complex system that is even more efficient at reproducing and perhaps more robust: able to survive and reproduce in environmental niches that the individual cells could never occupy. And, of course, since these complex systems, these organisms, depend on many of the same raw materials, there is an ever-increasing pressure to not only improve their own ability to acquire these resources, but to somehow prevent other organisms from acquiring them, or, better yet, just go ahead and take the resources from the other organisms. And then you have an arms race. Predators and prey. But also symbiosis and altruism. Life in all its complexity and variety. It is the elegance and beauty of the natural evolution of life that resonates most strongly with the evolution of Crichton's swarms.

Which is where I'll leave this discourse, except to note that while Crichton's novel *Prey* draws on some of the most exciting and profound areas of scientific research in the world today, and I'm perfectly happy to let him get away with some inaccuracies in order to get on with the story, those scientific details do matter out here in the real world and the scenario he describes is really not one you should lose any sleep over. Worry about "gray lung." Worry about corporate greed. Worry about personal and political irresponsibility. But flesh-eating nanoswarms you can pretty much lump in with brain-eating zombies and get on with your life. Does make for a good read, though.

LARRY YAEGER uses computers to solve hard and interesting problems, from flow fields over the space shuttle to special effects for *The Last Starfighter*, *2010*, and *Labyrinth*; to providing a computer "voice" for Koko the gorilla; to developing the world's first usable handwriting recognizer for second-generation Newtons and Mac OS X's "Inkwell"; to evolving machine intelligence in an Artificial Life system called Polyworld. He teaches programming and Artificial Life at Indiana University, and plays videogames and reads and watches entirely too much science fiction. And is eternally grateful to Levi for putting up with him! http://pobox.com/~larryy/.

References

Langton, C. G. *Artificial Life: The Proceedings of an Interdisciplinary Workshop on the Synthesis and Simulation of Living Systems*. Santa Fe Institute Studies in the Sciences of Complexity Proc. Vol. VI. Ed. C. Langton. Redwood City, CA: Addison Wesley, 1989.

Butler, S. (1880). *Unconscious Memory*. Project Gutenberg ebook http://www.gutenberg.org/etext/6605. Transcribed from the 1910 A. C. Fifield edition.

Waddington, C. H. *The Nature of Life*. London: Allen & Unwin, 1961.

Hillis, D. W. "Intelligence as an Emergent Behavior." *Daedalus, Journal of the American Academy of Arts and Sciences*, special issue on Artificial Intelligence. Winter (1988): 175–189.

Dennett, D. C. *Kinds of Minds: Toward an Understanding of Consciousness*. New York: Basic Books, 1996.

BE AFRAID. BE VERY AFRAID: MICHAEL CRICHTON'S *STATE OF FEAR*

David M. Lawrence

If there's anything that Michael Crichton makes clear in State of Fear *it is that the factors influencing climate change—and climate change research—are extremely complex at the very least. I'm afraid that it's going to take somebody like David Lawrence to even begin sort it all out for us.*

Now we are engaged in a great new theory, that once again has drawn the support of politicians, scientists, and celebrities around the world. Once again, the theory is promoted by major foundations. Once again, the research is carried out by prestigious universities. Once again, legislation is passed and social programs are urged in its name. Once again, critics are few and harshly dealt with.

Once again, the measures being urged have little basis in fact or science....

—MICHAEL CRICHTON, *State of Fear*

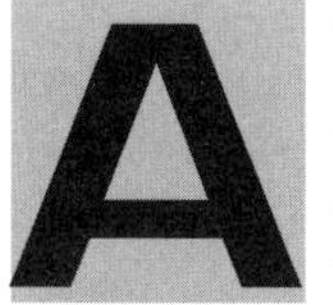

T LEAST AS FAR BACK as *The Andromeda Strain*, Michael Crichton has revealed in his writings skepticism about the limitation of science and technology as a tool in humanity's efforts to stave off disaster. Hubris and ignorance have led to the downfall of more than one of Crichton's protagonists, whether they be eaten by dinosaurs of their own creation or trapped in a lab with a deadly and spreading disease that they have unwittingly released. Often, such skepticism is warranted. Scientists are just as fallible as any other human, making mistakes large and small. Some mistakes lead to great disasters, such as the thalidomide scandal of the 1950s and 1960s, in which an inadequately tested medicine was, because of its efficacy in mitigating the effects of morning sickness, administered to the worst possible pool of patients: pregnant women. The problem, undiscovered until too late, was that thalidomide could cause severe birth defects in their children. Thousands of thalidomide babies, many born with shortened, even missing limbs, were the legacy of this failure by the scientific community.

That science can go wrong is no secret. The theme has been a staple of science fiction since the birth of the genre in the nineteenth century. The classic scientist-villain in these stories is usually evil, demented, or brilliant yet clueless, working alone or within a small organization, and almost always working beyond the fringes of the mainstream science of the time.

In *State of Fear*, Crichton takes this paranoia of science, and scientists, to new levels.

The book begins with an apparently authentic introduction by "MC" about a lawsuit to be filed on behalf of a small Pacific island nation, Vanuatu, against the U.S. Environmental Protection Agency for its failure to prevent global warming, which will apparently endanger the small nation through rising sea levels which flood the residents out of their homeland. Intrigue quickly follows, with a murder in Paris, a mysterious purchase in Malaysia, another killing in London, and mention of a radical environmental cause. The cause? Global warming, of course.

Global warming is an oft-used phrase. It, along with its lexicological cousins, climate change and the greenhouse effect, is blamed for many problems affecting human and natural systems. Many believe that the *tres amigos* will be the source of much mischief in the decades and centuries—even millennia—to come.

Despite Crichton's claim in an appendix to *State of Fear* that there is little basis for concern in fact or science, the existence of and mechanisms behind global warming, i.e., the greenhouse effect, are pretty established science. It was first described by the French mathematician Jean Baptiste Joseph Fourier in 1827.[1] The Swedish chemist Svante Arrhenius[2] measured the heat-trapping ability of carbon dioxide (or car-

[1] Jean Baptiste Joseph Fourier, "Mémoire sur les températures du globe terrestre et des espaces planétaires," *Mémoires de l'Académie royale des sciences de l'Institut de France* 7 (1827): 570–604.

[2] Svante Arrhenius, "On the Influence of Carbonic Acid in the Air upon the Temperature of the Ground," *Philosophical Magazine and Journal of Science*, 5th ser., 5. 41 (1869): 239–276.

bonic acid, as he called it) in a series of experiments he reported on in 1896. In fact, we would not be able to survive on the surface of our planet without it, for it is an important part of the radiation balance, which ultimately governs temperature, of the surface of the Earth.

Most of the energy that drives life and physical processes (such as photosynthesis, weather and atmospheric circulation, oceanic circulation, and physical and chemical weathering of soils) comes from the sun in the form of shortwave radiation—primarily visible and ultraviolet light. Some of that energy is scattered by molecules and particles in the atmosphere. Some is reflected back into space by clouds, for example, or by the surface. What is not reflected or scattered is absorbed. The molecules and materials that make up the atmosphere and surface of the Earth cannot absorb heat indefinitely. Some of that energy is used to do work, as in the coupling of carbon dioxide and water to make sugars via photosynthesis. What is not otherwise used, however, is given off as longwave radiation—infrared radiation, much of what we sense as heat. If that heat was allowed to pass freely back into space, the temperature at the surface of the Earth would be below freezing, about -19 degrees Celsius, or -2 degrees Fahrenheit. But the average surface temperature of the Earth is 14 degrees Celsius, or 57 degrees Fahrenheit. How can that be? [3]

The difference is the Earth's natural greenhouse effect. Gases in the atmosphere, such as water vapor (the most abundant), carbon dioxide (which, with water vapor, is an end product of the burning of fossil fuels), and methane (one of the most potent natural greenhouse gases), trap heat near the surface like a blanket, keeping the temperature about thirty-three degrees Celsius, or fifty-nine degrees Fahrenheit, warmer than otherwise possible. The Earth's two nearest planetary neighbors, Venus and Mars, serve as bookends, so to speak, on the influence of greenhouse gases on surface temperatures.

Though the Martian atmosphere is about 95 percent carbon dioxide, the atmosphere is thin, much more like a sheet than a blanket. While one would expect the surface temperatures of Mars to be some-

[3] Hervé Le Treut, Robert Somerville, Ulrich Cubasch, Yihui Ding, Cecilie Mauritzen, Abdalah Mokssit, Thomas Peterson, and Michael Prather, "Historical Overview of Climate Change," in *Climate Change 2007: The Physical Science Basis. Contribution of Working Group I to the Fourth Assessment Report of the Intergovernmental Panel on Climate Change*, eds. Susan Solomon, Dahe Qin, Martin Manning, Zhenlin Chen, Melinda Marquis, Kristen Averyt, Melinda M.B. Tignor, and Henry Leroy Miller Jr. (New York: Cambridge University Press, 2007), 93–127.

what cooler than that of Earth because of its increasing distance from the Sun, Mars is much cooler—about fifty degrees Celsius, or ninety degrees Fahrenheit, cooler than Earth. Mars was much warmer, with liquid water at the surface, but the planet apparently entered a reverse greenhouse effect: carbon dioxide was removed from the atmosphere, reacting with and binding to rocks at the surface. As the carbon dioxide was removed from the atmosphere, the gaseous envelope surrounding the planet thinned, temperatures dropped, and the other major greenhouse gas present, water vapor, froze, becoming ice on the surface. The loss of atmospheric water vapor further aggravated the cooling.[4,5]

Venus on paper, on the other hand, should have been the Earth's twin. But there are differences. The Earth, because it was farther from the sun, had somewhat cooler surface temperatures which allowed vast oceans of liquid water to cover the surface. These surface waters could dissolve carbon dioxide from the atmosphere. Life on the surface could use atmospheric carbon as biological building blocks. Venus was closer to the sun, therefore hotter because of the greater amounts of solar radiation it received. Oceans of liquid water either could not form, or, as its atmosphere warmed, more and more water evaporated from the surface. Greenhouse gases otherwise dissolved in the early Venusian oceans or bound in its surface rocks were released to the atmosphere as well. As the concentration of greenhouse gases increased, so did the temperatures, leading to further release of greenhouse gases into the atmosphere and further warming—in other words, a runaway greenhouse.[6,7] The surface temperature of Venus now averages about 460 degrees Celsius, or 860 degrees Fahrenheit.

Humans, by the combustion of fossil fuels such as coal and petroleum and by the conversion of natural landscapes to agricultural and urban uses, have triggered an increase in the concentration of several greenhouse gases in the atmosphere. The concentrations of carbon dioxide, methane, and nitrous oxide have increased markedly since the

[4] James F. Kasting, Owen B. Toon, and James B. Pollack, "How Climate Evolved on the Terrestrial Planets," *Scientific American* 256 (1988): 90–97.

[5] Donald M. Hunten, "Atmospheric Evolution of the Terrestrial Planets," *Science* 259 (1993): 915–920.

[6] Andrew P. Ingersoll, "The Runaway Greenhouse: A History of Water on Venus," *Journal of the Atmospheric Sciences* 26 (1969): 1191–1198.

[7] Hunten, "Atmospheric Evolution of the Terrestrial Planets."

beginning of the Industrial Revolution in 1750. Carbon dioxide has increased from a pre-industrial level of 280 ppm (parts per million) to 379 ppm in 2005. If current emission trends continue unabated, it will likely double pre-industrial levels by the end of this century.[8] Atmospheric methane has more than doubled, from 715 ppb (parts per billion) in pre-industrial times to 1774 ppb in 2005, although the growth rate in the methane concentration has decreased somewhat since the early 1990s.[9] The nitrous oxide concentration has risen from a pre-industrial level of 270 ppb to 310 ppb in 2005.[10] Data from ice cores suggest that the current levels of carbon dioxide and methane exceed anything seen in the last 650,000 years.[11]

The concern is that greenhouse gases will do as they are known to do: trap more heat near the surface of the Earth, therefore altering temperature patterns and triggering potentially catastrophic environmental changes. Many argue that significant changes in our behavior are required to stem the increase and stave off disaster. Some, including Crichton, argue otherwise.

In an appendix to *State of Fear*, Crichton compares the scientific consensus of concern over global warming to a number of scientific abuses during the twentieth century. One was eugenics, in which many sought to improve the quality of humanity by encouraging the breeding of desirables—essentially intelligent, wealthy, blueblood, "white" people, and discouraging or even preventing the breeding of undesirables. Undesirables included people of color (or of mixed race), so-called "white trash," homosexuals, petty criminals, and people considered mentally deficient. Many leaders and institutions in the United States promoted eugenics and conducted eugenics research. Adolf Hitler drew aid and comfort from what was happening in the United States, learning much of what he needed from America to implement

[8] Piers Forster, Venkatachalam Ramaswamy, Paolo Artaxo, Terje Berntsen, Richard Betts, David W. Fahey, James Haywood, Judith Lean, David C. Lowe, Gunnar Myhre, John Nganga, Ronald Prinn, Graciela Raga, Michael Schulz, and Robert Van Dorland, "Changes in Atmospheric Constituents and in Radiative Forcing," in *Climate Change 2007*, 129–234.

[9] Ibid.

[10] Ibid.

[11] Eystein Jansen, Jonathan Overpeck, Keith R. Briffa, Jean-Claude Duplessy, Fortunat Joos, Valérie Masson-Delmotte, Daniel Olago, Bette Otto-Bliesner, W. Richard Peltier, Stefan Rahmstorf, Rengaswamy Ramesh, Dominique Raynaud, David Rind, Olga Solomina, Ricardo Villalba, and De er Zhang, "Paleoclimate," in *Climate Change 2007*, 433–497.

the Holocaust. (There was "scientific" cooperation between America and the Nazis prior to the onset of World War II.)

Crichton draws another cautionary lesson from Josef Stalin's Soviet Union. Trofim Denisovich Lysenko, an agricultural scientist who rejected the developments of modern genetics and evolution in favor of the old, discredited theory of Larmarckism—inheritance of acquired characteristics—promised increased agricultural yields without fertilizing fields. He promoted a process called vernalization that was purported to improve flowering of crops in spring by exposing the seeds to prolonged cold in the winter. Such treatment does increase flowering in some crops, but Lysenko took the idea a step further, claiming that the descendants of treated individuals would inherit the increased ability to flower without having to undergo the cold treatment. This became known as Lysenkoism. His ideas were a godsend to a Soviet Union reeling from famines in which millions died, for they promised far greater crop yields without a corresponding increase in investment. The problem was they did not work.

Eugenics is offered as a warning against social movements sold as scientific programs. Lysenkoism is offered as a warning against the politicization of science. Crichton believes that both phenomena lie at the heart of the concern over global warning. It is from this point of view that *State of Fear* is written.

Crichton expresses most of his skepticism through the voice of one character, John Kenner, a Massachusetts Institution of Technology professor-cum-secret agent—a man just as lethal, but much better educated, than Ian Fleming's literary (not celluloid) James Bond. The philosophical aspect of Kenner seems to be based on a living MIT professor, Richard S. Lindzen, who is a prominent global warming skeptic. He is not, so far as I know, an intelligence agent. But the secret agent aspect is not that farfetched, as academics are known to work overtly or covertly for intelligence agencies.

Crichton doesn't wait for Kenner to appear in the book before taking his first shot at the current concern over global warming. The setting for the shot is, appropriately, Iceland, where George Morton, a wealthy backer of environmental causes, Peter Evans, Morton's attorney and chief protagonist, and Nicholas Drake, head of the National Environmental Resource Fund (NERF) and chief villain, visit a glaciologist working on a project supported by NERF with the help of Morton's money. While Morton and Evans are being distracted by the local scenery (in the form of beautiful Icelandic geologists), Drake and the principal investigator are arguing about the researcher's findings: that temperatures are cooler in Iceland at the time the novel takes place (2004) than they had been early in the twentieth century; and that the glaciers, which had receded during the earlier warm period, were now surging. The researcher wants to publish his results without obfuscation; Drake, the "environmentalist," wants the facts withheld so as not to confuse the public over the inevitability and seriousness of the oncoming global catastrophe.

It is at this point that Crichton introduces the first of many references to actual scientific literature to bolster his argument that concern over global warming is overblown: this first offering is the paper "Global Warming and the Greenland Ice Sheet," published in the journal *Climatic Change* in 2004. In his footnote, he quotes the article, "Since 1940 . . . data have undergone predominantly a cooling trend. . . . The Greenland ice sheet and coastal regions are not following the current global warming trend." [12] All this appears damning, but this barely scratches the surface of the article; the quote Crichton selected is actually from the abstract, not the more meaty discussions of the research in the body of the text.

The lead author of the study, Petr Chylek, now of Los Alamos National Labs, is often listed as a global warming skeptic. He is on record saying there is insufficient evidence to link climate conditions today with global warming. Nevertheless, nowhere in this article does he say his findings should be used to discount current concerns. The article points out something that all climate scientists know: there is considerable local variation in weather and climate. The growth or decline of glaciers derives from a complex balance of temperature and moisture ability. Warmer temperatures do melt ice, but warmer temperatures may also bring more precipitation—warm air holds more water vapor, which can be transported far from the source to increase rain or snowfall elsewhere. If more ice is lost through melting than is gained through precipitation, the glaciers shrink. If more ice

[12] Petr Chylek, Jason E. Box, and Glen Lesins, "Global Warming and the Greenland Ice Sheet," *Climatic Change* 63 (2004): 201–224.

is gained through precipitation than is lost through melting, the glaciers grow. The nature of the balance can lead to perverse effects: glaciers can shrink during cooler times and grow during warmer times.

In Greenland's case, Chylek and his colleagues suggested that Greenland is strongly affected by the North Atlantic's version of the notorious El Niño, the North Atlantic Oscillation (NAO). El Niño is associated with a fluctuating atmospheric pressure pattern in the Pacific known as the Southern Oscillation. Normally, atmospheric pressure is higher in the eastern Pacific (off Ecuador) and lower in the western Pacific (in the neighborhood of Australia). As a result, strong tropical winds blow from east to west; arid conditions prevail in the eastern Pacific and humid conditions prevail in the west. During an El Niño, the pressure and wind patterns reverse, triggering weather anomalies that can have catastrophic effects around the globe.

The North Atlantic Oscillation is a similar pressure fluctuation between a region of typically high pressure over the Azores and a region of typically low pressure in the neighborhood of Iceland. The pressure differences between these two locations affect the mid-latitude westerly winds blowing across the Atlantic. When the pressure differences are high, strong westerly winds bring stronger, more frequent storms to Europe in winter. As a result, Europe has warmer, wetter winters, as does the eastern United States. Canada and Greenland, however, have colder and drier winters. When the pressure differences are low, weaker westerlies lead to fewer and weaker winter storms in Europe. Northern Europe experiences colder conditions, southern Europe experiences humid conditions. Outbreaks of cold air sweep over the eastern United States, bringing more frequent snowstorms. Weather conditions over Greenland, however, are milder.

By now, it should be clear that the North Atlantic Oscillation has a tremendous influence on Greenland's weather, therefore it has a tremendous influence on Greenland's glaciers and may even counteract the effect of global warming. Subsequent studies by Chylek have borne this out. Despite Chylek's skepticism about global warming, he was the lead author of a study published in the journal *Geophysical Research Letters* in 2005 that supports the concern over global warming. The study, written with Ulrike Lohmann of the Swiss Federal Institute of Technology, focused on northeastern Greenland, a portion not affected by the North Atlantic Oscillation. The two scientists found late twentieth-century *warming*, not cooling, that is consistent with global warming predictions.[13] In fact, temperatures in that part of Greenland are rising twice as fast as in the rest of the globe! The last sentence in the paper's concluding section says, "Our analysis suggests an agreement between observation and climate model predictions of the rate of temperature change due to global warming in Greenland and its ratio to the rate of global temperature change."

Kenner addresses the questions of ice sheets and glaciers in other parts of the world in at least two other places in the book. One of these passages, accompanied by nine references from the scientific literature, addresses whether or not Antarctica is melting. Much of Antarctica is not melting—no serious climate scientist expects the ice mass in the interior of the vast southern continent to do so. Antarctica is isolated from other continents by the Southern Ocean, a vast, cold body of water accompanied by weather systems that acts as a chiller—the cold waters absorb heat from southward-moving air masses that pass over them en route to the southern pole. The interior of the continent is a vast, high plateau. In the troposphere—the lower layer of the atmosphere in which almost all "weather" occurs—the higher you go, the colder it gets. Thus, the high elevations of Antarctic interior likewise serve to keep temperatures frigid. Thus, there's little reason to expect much, if any, warming in the Antarctic interior.

But there are data to suggest that parts of Antarctica are cooling—this is the evidence that Crichton highlights to dispel notions of any real global warming. The problem with Crichton's argument is that the data series that show this cooling are of fairly short duration—most are series of less than fifty years—way too short to draw any statistically sound conclusions about trends. Another problem is that, while a slight majority of the continent appears to be cooling—about 60 percent according to Peter Doran of the University of Chicago, the author of one of the papers Crichton cites—the rest is, well, warming.[14] One

[13] Petr Chylek and Ulrike Lohmann, "Ratio of the Greenland to Global Temperature Change: Comparison of Observations and Climate Modeling Results," *Geophysical Research Letters* 32 (2005): L14705, doi:10.1029/2005GL023552.

[14] Peter T. Doran, John C. Priscu, W. Berry Lyons, John E. Walsh, Andrew G. Fountain, Diane M. McKnight, Daryl L. Moorhead, Ross A. Virginia, Diana H. Wall, Gary D. Clow, Christian H. Fritsen, Christopher P. McKay, and Andrew N. Parsons, "Antarctic Climate Cooling and Terrestrial Ecosystem Response," *Nature* 415 (2002): 517–520.

of the areas that is warming, the Antarctic Peninsula, a fingerlike projection that points north toward the tip of South America, is warming at a far higher rate than the rest of the planet. Several large ice sheets that used to cling to the edges of Antarctica, the Larsen A, Larsen B, and the Wilkins ice shelves, each have collapsed suddenly in the last fifteen years. Warmer temperatures overall, longer melt seasons, and the destabilizing effects of surface meltwater as it seeps into the ice below have contributed to the disintegration of these massive accumulations of ice. The Larsen B ice shelf faced a double whammy: warm air temperatures and meltwater eating at it from above and warm currents eating at it from below. Most of it broke up in a matter of days.

Crichton mentions a 1999 study in the journal *Nature* that found that maximum temperatures during the last four interglacials—warm periods in between the ice ages—were warmer than today.[15] That may be true, but the last four interglacials are long since over. Crichton fails to note the fallacy of comparing an ongoing event to similar events that have run their course. None of the four previous interglacials can get any warmer; they are all finished. The current warm period, called the Holocene by earth scientists, has not yet run its course. No one will *know* whether or not it ends up warmer, colder, or about the same as the previous four interglacials for several hundred, or even several thousand, more years.

It might not be wise to wait until the year 3000 to make sure the current warm period is hotter than its predecessors before taking action to combat global warming.

A number of studies have found that the glacial (ice age)/interglacial cycles are closely related to characteristics of the Earth's orbit around the sun as well as the Earth's tilt on its axis. The two factors largely control the amount and distribution of solar radiation that strikes the Earth. Few climate scientists would challenge that statement today. Nevertheless, there is considerable room for the influence of greenhouse gases. The 1999 study cited by Crichton in an effort to cast doubt upon the concept of global warming instead finds greenhouse gases important.

[15] J. R. Petit, J. Jouzel, D. Raynaud, N. I. Barkov, J. M. Barnola, I. Basile, M. Bender, J. Chappellaz, M. Davisk, G. Delaygue, M. Delmotte, V. M. Kotlyakov, M. Legrand, V. Y. Lipenkov, C. Lorius, L. Pépin, C. Ritz, E. Saltzmank, and M. Stievenard, "Climate and Atmospheric History of the Past 420,000 Years from the Vostok Ice Core, Antarctica," *Nature* 399 (1999): 429–436.

> These results suggest that the same sequence of climate forcing operated during each termination [of an ice age]: orbital forcing (with a possible contribution of local insolation changes) followed by two strong amplifiers, greenhouse gases acting first, then deglaciation and ice-albedo feedback.

The final sentences of the 1999 study go even further to remind readers that it is premature to be as dismissive as Crichton is of the threat of global warming.

> Finally, CO_2 and CH_4 concentrations are strongly correlated with Antarctic temperatures; this is because, overall, our results support the idea that greenhouse gases have contributed significantly to the glacial-interglacial change. *This correlation, together with the uniquely elevated concentrations of these gases today, is of relevance with respect to the continuing debate on the future of Earth's climate.* (emphasis added)

Later in the book, Crichton engineers a scene where Kenner the MIT professor engages a character named Ted Bradley, a Hollywood actor active in environmental causes such as those espoused by NERF, in a rather uneven duel over the scientific evidence for or against global warming. Bradley gets flustered, at one point muttering "all the glaciers melting" in a list of warning signs of global warming. Kenner twists the statement so that it seems those concerned with global warming believe literally that all glaciers are melting. (Kenner does this twice in less than one page of text.) No one who is properly informed—not even Al Gore—believes all glaciers are melting. But this piece of literary trickery implies such, casting doubt on the sanity and/or scientific competence of those concerned about global warming.

Kenner concedes that some glaciers are shrinking, while others are not. But he presses his argument further: No one knows whether the majority of glaciers are getting smaller. Then he says there is no way we can know: detailed mass balance data (measures of the amount of ice that accumulates via precipitation versus that lost through melting or other processes) are available for only a small number of glaciers worldwide. This latter point sounds convincing, but it's got a major problem. There isn't a single field across the entire spectrum of academic disciplines in which a large percentage of the population of interest has been scientifically sampled. Everything scientists know in any discipline in which scientists are involved is based on the analysis of a small subset of the whole. In order to damn the work of those who study glaciers, Crichton damns all sciences.

Crichton accurately quotes Roger J. Braithwaite, who wrote a review article in *Progress in Physical Geography* on the status of glacier mass balance studies in the latter part of the twentieth century, that "There is no obvious common global trend of increasing glacial melt in recent years."[16] The time period analyzed by Braithwaite ended in 1995. His criticisms were that most records were too short (generally less than ten years) to draw reliable conclusions; that there was a lack of adequate representation of glaciers from regions outside of North America, Europe, and the former Soviet Union; that most glaciers analyzed were from moist, maritime environments rather than from the dry, cold environments characteristic of many alpine glaciers; and that the methods traditionally used to estimate mass balance were fraught with error—the errors stemming from difficult field conditions and the complicated nature of the environments in which the glaciers are found.

Many of the weaknesses cited by Braithwaite have since been addressed. More glaciers in the Andes and Patagonia, the Eurasian Arctic, the mountains of central and southern Asia, and the Sub-Antarctic islands have been studied, thus improving the global coverage of mass-balance analyses. Improved methods have been applied and ways to reduce errors inherent in traditional methods of obtaining mass balance data. Short records have been lengthened by additional data.

With this new and improved data, it is reasonable to conclude that glaciers in many parts of the world are shrinking. According to the National Snow and Ice Data Center (NSIDC), which uses satellite data instead of the traditional field-based methods to obtain mass balance data, large volumes of ice have been lost from glaciers in Alaska, northwestern United States, southwestern Canada, the mountain spine of Asia, and Patagonia. The findings of the NSIDC project are supported by those of other glacier studies using other—including traditional—methods. The most recent revision to the Glacier Mass Balance and Regime database, compiled by Mark Dyurgerov of the University of Colorado's Institute of

[16] Roger J. Braithwaite, "Glacier Mass Balance: The First 50 Years of International Monitoring," *Progress in Physical Geography* 26 (2002): 76–95.

Arctic and Alpine Research, lists traditionally derived mass balance data for 304 glaciers worldwide, including some from areas originally listed as underrepresented in the Braithwaite review, over a collective period from 1946 to 2003. The data are somewhat difficult to compare because of the variation in lengths of the samples. About one-third of the dataset consists of a series of less than five years; of those, forty-five series contain only one year of measurements. A series of more than forty years in length makes up one-tenth of the dataset; the longest series spans fifty-eight years. Of those data series with more than ten years of measurements, 102 glaciers had a net negative mass balance (loss of ice); only fifteen had a net positive mass balance. When a series of three or more years in length is analyzed, 185 have a net negative mass balance; only forty-nine have a net positive mass balance.[17] The trends in both series are similar. Ice mass losses averaged about 290 mm/year in equivalent water depth—the way precipitation amounts are measured—from 1951 through 1955, increasing to just over 300 mm/year during the next five-year period. Ice mass losses decreased to about 80 mm/year during 1971 through 1975. Losses have steadily increased since, to about 500 mm/year from 1996 through 2000. The years 2001 through 2003 (the last year for which sufficient data are available) were even higher, averaging about 1000 mm/year.[18] The regions in which ice mass losses have occurred are widespread: North America, much of Eurasia (including Europe, the former Soviet Union, and South Asia), Iceland, Kenya, South America (including Patagonia), New Zealand, and some of the Sub-Antarctic islands.[19]

Temperature decreases with altitude in the troposphere. This temperature gradient can affect the local mass balance on a glacier. In the upper portion, cooler temperatures may lead to an accumulation (net mass gain) of ice. In the lower portion, warmer temperatures may lead to a net mass loss of ice. The elevation where the balance is zero (no net gain or loss over the course of a year) is the equilibrium-line altitude. During warmer climate phases, the equilibrium-line altitude will

[17] Mark Dyurgerov, *Glacier Mass Balance and Regime Measurements and Analysis*, 1945–2003, eds. Mark Meier and Richard Armstrong. (Boulder, Colo.: Institute of Arctic and Alpine Research, University of Colorado; distributed by the National Snow and Ice Data Center, Boulder, CO, http://nsidc.org/data/g10002.html, accessed 30 Aug. 2007).

[18] Ibid.

[19] Ibid.

be higher. During cooler phases, it will be lower. Dyurgerov reported in 2002 that the equilibrium line has risen globally by about 200 meters in the latter half of the twentieth century.[20]

Kenner is closer to the truth when he addresses one of the poster children of global warming: the shrinking snows of Mount Kilimanjaro. Kilimanjaro is a massive volcano located near the equator in Tanzania. For as far back as anyone can remember, its summit has been covered with snow and ice. But the glaciers have been receding since the late 1800s. The decline continues, although the pace of the decline is much reduced, today. Despite the imagery depicting the shrinking glaciers of Kilimanjaro in discussions of global warming, however, global warming per se may have little to do with it. For one, the glaciers began receding decades before the effects of global warming were noticeable. While there is evidence of a slight warming at lower elevations, there is no evidence of warming at the level of the summit—in part because no long-term temperature measurements exist. Satellite measurements of the temperature of the upper part of the troposphere, balloon-based measurements, and computer models all indicate little or no warming in the last few decades in the elevation band where Kilimanjaro's glaciers are located.[21,22] While these data are suggestive, they do not constitute proof. Nevertheless, it is reasonable to conclude that temperature changes have little to do directly with the loss of Kilimanjaro's ice cap.

What has changed? Land use surrounding the mountain, for one. The clearing of forests for human uses has altered the local climate regime, resulting in a reduction of precipitation. Trees typically pump a lot of water vapor back into the atmosphere via a process called transpiration. The vegetation that has replaced the forests—grasses and agricultural crops—does not transpire as much as trees. As the atmospheric moisture source dries up, precipitation goes down. Georg Kaser, a scientist at the University of Innsbruck, has suggested that such changes have altered the mass balance of ice at Kilimanjaro's sum-

[20] Mark Dyurgerov, *Glacier Mass Balance and Regime: Data of Measurements and Analysis*, eds. Mark Meier and Richard Armstrong. Institute of Arctic and Alpine Research, Occasional Paper No. 55 (Boulder, Colo.: Institute of Arctic and Alpine Research, University of Colorado, 2002).

[21] Georg Kaser, Douglas R. Hardy, Thomas Mölg, Raymond S. Bradley, and Tharsis M. Myera, "Modern Glacier Retreat on Kilimanjaro as Evidence of Climate Change: Observations and Facts," *International Journal of Climatology* 24 (2004): 329–339.

[22] Philip W. Mote and Georg Kaser, "The Shrinking Glaciers of Kilimanjaro: Can Global Warming Be Blamed?" *American Scientist* 95 (2007): 318–325.

mit. Most of the ice is lost through sublimation, a process by which water changes state from ice directly into gas (water vapor) without passing through a liquid state. The energy to drive this sublimation is provided primarily by solar radiation (shortwave radiation), not the longer wavelengths (infrared) that we sense as temperature. Even if the amount of shortwave energy striking the ice remains constant, a reduction in moisture supply will lead to a loss of ice mass if all the water lost through sublimation cannot be replaced by precipitation.[23] While the argument that land use changes and resulting depletion of the moisture supply are the primary causes of the shrinking of Kilimanjaro's glaciers is persuasive, the researchers consistently concede that global warming may have an indirect effect as climate fluctuations in the region and elsewhere also affect moisture supply. "There is strong evidence of an association over the past 200 years or so between Indian Ocean surface temperatures and the atmospheric circulation and precipitation patterns that either feed or starve the ice on Kilimanjaro," wrote Philip Mote and Georg Kaser in an article in *Scientific American* in 2007.[24]

Throughout *State of Fear*, Crichton presents graphics of temperature trends from sites that show cooling instead of warming. In one particularly long section, he serves an excellent educational purpose in that he shows how, by selective presentation, the same set of data can lead an observer to opposite conclusions. (A better primer on statistical tricks, however, is Darrell Huff's 1954 classic, *How to Lie with Statistics*.) Crichton's purpose with the graphics is less to educate than to cast doubt on the concept of global warming triggered by combustion of fossil fuels. How can such warming be occurring if he can readily find data from sites that show a cooling trend instead?

Crichton's trick, unfortunately, is just that: a trick. If the global average temperature is increasing, it is increasing. Period.

To use a sports analogy, consider the average height of the 1987–1988 Washington Bullets, with one future hall-of-famer on the roster, Moses Malone, and another future hall-of-famer, Wes Unseld, who was to take over in mid-season as coach. The Bullets averaged six feet,

[23] Thomas Mölg and Douglas R. Hardy, "Ablation and Associated Energy Balance of a Horizontal Glacier Surface on Kilimanjaro," *Journal of Geophysical Research* 109 (2004): D16104, doi:10.1029/2003JD004338.

[24] Mote and Kaser, op cit.

six inches in height. Despite the fact that the roster also featured Tyrone Curtis "Muggsy" Bogues, at a towering five feet, three inches, and at the other end of the scale Manute Bol, at seven feet, six inches, the team still had an average height of six feet, six inches. Both Bogues and Bol were rather significant outliers—exceptions, if you will—but their presence did not make the math any less valid.

In this passage, Crichton performs another piece of intellectual sleight of hand in an exchange of dialog between Evans, the chief protagonist, and Jennifer Haynes, an attorney working for NERF who also happens to be Kenner's niece. (Character identifications added for clarity.)

HAYNES: "So, according to the theory, the atmosphere itself gets warmer, just as it would inside a greenhouse."

EVANS: "Yes."

HAYNES: "And these greenhouse gases affect the entire planet."

EVANS: "Yes."

HAYNES: "And we know that carbon dioxide—the gas we all worry about—has increased the same amount everywhere in the world. . . ."

(DML: At this point, Haynes pulls out the Keeling curve showing measured atmospheric levels of carbon dioxide from 1957 through 2002. It shows rising levels of the gas.)

EVANS: "Yes. . . ."

HAYNES: "And its effect is presumably the same everywhere in the world. That's why it's called *global* warming."

For Crichton and Haynes, the case is closed at this point. Evans goes on to mount a feeble defense, saying that he's heard that according to global warming theory, some places may get colder even as the planet warms. Haynes pursues the attack, recalling the temperature records of Albany and New York City. Albany appears to be cooling about 0.25 degrees Celsius or 0.5 degrees Fahrenheit from 1820 through 2000. New York City is warming, about three degrees Celsius or five degrees Fahrenheit from 1822 through 2000. West Point, New York, roughly halfway between Albany and New York City, shows little or no temperature trend from 1826–2000. The distance between Albany and New

York City is only about 230 kilometers. Crichton, through Haynes, asks, is it logical to expect so much variation in so short a space? Or is it evidence that temperature measurements are capturing something other than global warming?

The answer to the first question is yes. It is logical to expect such variation in so short a space. The New York State Climate office puts it plainly:

> The climate of New York State is broadly representative of the humid continental type, which prevails in the northeastern United States, but its diversity is not usually encountered within an area of comparable size. The geographical position of the state and the usual course of air masses, governed by the large-scale patterns of atmospheric circulation, provide general climatic controls. Differences in latitude, character of the topography, and proximity to large bodies of water have pronounced effects on the climate.[25]

Despite the rather small distance between Albany and New York City, there are substantial differences in climate. Albany is in the Upper Hudson River Valley. New York City is at the mouth of the Hudson River, along the Atlantic Coast. Cold, dry air masses blow into New York from the northern interior of North America. Warm, moist air travels up from the Gulf of Mexico, Caribbean, and tropical North Atlantic. Cool maritime air travels into the region from adjacent portions of the North Atlantic. Albany more often receives the cold, dry air masses. New York City more often receives the subtropical and maritime air masses.[26]

This is reflected in climate data from the two cities. The average annual temperature between Albany and New York City (Central Park) differs by 3.9 degrees Celsius or 7.0 degrees Fahrenheit—with Albany, obviously, being cooler (8.7 degrees Celsius/47.6 degrees Fahrenheit versus 12.6 degrees Celsius/54.6 degrees Fahrenheit). Low January temperature in Albany is 7.2 degrees Celsius or 12.9 degrees Fahrenheit cooler than in New York City (-10.4 degrees Celsius/13.3 degrees Fahrenheit versus -3.2 degrees Celsius/26.2 degrees Fahrenheit). In the summer, however,

[25] A. Boyd Pack, "Climates of the States: New York," in *Climatography of the United States*, No. 60–30. (Washington, D.C.: U.S. Department of Commerce, 1973).

[26] Ibid.

temperature differences are minimal: high July temperature in Albany is only 1.1 degrees Celsius or 2.0 degrees Fahrenheit cooler than in New York City (27.9 degrees Celsius/82.2 degrees Fahrenheit versus 29.0 degrees Celsius/84.2 degrees Fahrenheit). New York City gets more annual precipitation, 1262 mm or 49.7 inches versus 980 mm or 38.6 inches in Albany.[27] Albany, on the other hand, gets more of its precipitation in the form of snow: 1598 mm or 62.9 inches (snow depth, not equivalent water depth) versus 566 mm or 22.3 inches in New York City.[28,29] Furthermore, Albany has recorded below-freezing temperatures in every month except July and August. New York City, on the other hand, has recorded below-freezing temperatures in only five months: January, February, March, November, and December.[30,31]

In truth, Albany and New York City are part of the same climate regime. The march of the seasons follows the same general pattern—hot summers, cold winters, plenty of precipitation year-round—in both places. Despite that, and despite what is said in *State of Fear* about the cities' proximity, the two cities have enough differences in the frequency and types of weather systems that affect them to explain why Albany is experiencing a cooling trend, why New York City is experiencing a warming trend, and why West Point, located somewhere in between, has a climate pattern that is likewise somewhat in between the other two cities. In climate, there is no simple linear process—If A, then B—that ensures a uniform response to global warming everywhere.

With respect to the second question—are temperature measurements capturing something other than global warming?—the answer is yes. To an extent. The culprit is the urban heat island effect. The effect, in which the temperature of an urban area is generally higher than that of its non-urban surroundings, is well known; it was first de-

[27] U.S. Department of Commerce. National Oceanic and Atmospheric Administration. *Monthly Station Normals of Temperature, Precipitation, and Heating and Cooling Degree Days 1971–2000: New York*. Climatography of the United States No. 81. (Asheville, N.C.: U.S. Department of Commerce, National Oceanic and Atmospheric Administration, National Climatic Data Center, 2002.)

[28] New York State Climate Office. *Station Summary: Albany*. (Ithaca, N.Y.: New York State Climate Office, http://nysc.eas.cornell.edu/albany_c20.html, accessed 5 Sep. 2007.)

[29] New York State Climate Office. *Station Summary: New York*. (Ithaca, N.Y.: New York State Climate Office, http://nysc.eas.cornell.edu/newyork_c20.html, accessed 5 Sep. 2007.)

[30] New York State Climate Office. *Station Summary: Albany*.

[31] New York State Climate Office. *Station Summary: New York*.

scribed in 1833.[32] The urban heat island effect can be a problem for global warming studies because of where much of the data documenting temperature change comes from: long-term temperature records obtained from weather stations around the globe. Many, though not all, weather stations are based in urban areas. The globe is becoming increasingly urbanized, with humans around the world abandoning rural areas in the hopes of a better life in the cities. This trend is greatest in developing nations, where cities are expanding in order to accommodate the burgeoning populations. As urban areas expand, more or less natural landscapes are replaced by landscapes of pavement and buildings. In developed nations, the growth of suburbs around the urban core leads to further loss of non-urban environments.

The temperature differences between urban and non-urban areas stem from a number of factors. Natural landscapes and developed landscapes differ in their heating and cooling characteristics. Urban landscapes seem to store more energy over the course of the day and release it throughout the night, leading to the most noticeable effect of urban heat islands: warmer nighttime temperatures than in non-urban areas. (Effects on daytime temperatures are relatively minor.) In addition, non-urban landscapes tend to store water in the vegetation and soil. Urban landscapes, because of their impervious surfaces, store little moisture either above or below ground—precipitation and snowmelt run off along the surface into streams instead. Where moisture is available, solar energy is consumed by evaporation and transpiration (water loss through the leaves of plants). The energy is essentially stored in water vapor molecules in the atmosphere rather than used to heat the surrounding environment. In urban environments where surface moisture is lacking, that energy is available for surface heating instead. A significant amount of heat is generated by human activities as well; an air conditioner may cool the inside of a building, but that removed from the inside is vented outside, for example. Urban environments affect regional climate in other ways, for example, by altering patterns of wind flow and by serving as a source for aerosols and other particles that can have climatic effects. Urban heat island effects are greatest in winter as well as at times when the winds are light.

[32] Luke Howard, *The Climate of London, Deduced from Meteorological Observations, Made in the Metropolis and at Various Places Around It*, 2nd ed. 3 vols. (London: Harvey and Darton, 1833).

When air temperatures are taken in urban areas, the raw temperature measurements reflect the urban heat island effect. In order to detect global warming from greenhouse gases, then, the effect must be statistically filtered out of the temperature signal. No matter how such filtering is done, the results are imperfect. The only perfect method to filter out the urban heat island effect is to construct a parallel universe with an Earth identical in all respects except for a lack of urban areas so that scientists could compare temperature measurements from both. As this method seems somewhat infeasible, statistical techniques are the best option. The most common of the statistical methods apply a filter based on population size as an indication of urbanization. Crichton, in a footnote, accurately quotes one review article that addresses the topic, "Recent studies suggest that attempts to remove the 'urban bias' from long-term climate records (and hence identify the magnitude of the enhanced greenhouse effect) may be overly simplistic."[33] This is a point that few climate scientists would dispute.

A number of studies have indicated that population-based methods may underestimate the magnitude of the urban heat island effect.[34,35] On the other hand, a number of studies suggest that the urban heat island effect, no matter how real it may be, contributes little to the global warming signal evident in surface temperature trends as well as a number of other types of data.[36,37,38] Because the urban heat island effect is most noticeable at night, and because it is greater on calm nights than on windy ones, David E. Parker of the Hadley Centre in the United Kingdom devised a test to detect the urban heat island effect by comparing nighttime minimum temperatures on calm versus windy nights. No difference was detected, which indicated that much of the climate warming during the

---

[33] Ian G. McKendry, "Applied Climatology," *Progress in Physical Geography* 27 (2002): 597–606.

[34] Stanley A. Changnon, "A Rare Long Record of Deep Soil Temperatures Defines Temporal Temperature Changes and an Urban Heat Island," *Climatic Change* 42 (1999): 531–538.

[35] Ross McKitrick and Patrick J. Michaels, "A Test of Corrections for Extraneous Signals in Gridded Surface Temperature Data," *Climate Research* 26 (2004): 159–273.

[36] P. D. Jones, P. Ya. Groisman, M. Coughlan, N. Plummer, W. C. Wang, and T. Karl, "Assessment of Urbanization Effects in Time Series of Surface Air Temperatures Over Land," *Nature* 347 (1990): 169–172.

[37] Thomas C. Peterson, "Assessment of Urban Versus Rural In Situ Surface Temperatures in the Contiguous United States: No Difference Found," *Journal of Climate* 16 (2003): 2941–2959.

[38] J. Hansen, R. Ruedy, M. Sato, M. Imhoff, W. Lawrence, D. Easterling, T. Peterson, and T. Karl, "A Closer Look at United States and Global Surface Temperature Change," *Journal of Geophysical Research* 106 (2001): 23947–23963.

past century was due to some other factor.[39] In some instances, nighttime minimum temperatures were warmer on windy nights than on calm ones—opposite what is expected of an urban heat island.[40] Parker's findings have not gone without challenge.[41]

Despite the controversy over the urban heat island effect, a group of scientists working under the auspices of the Intergovernmental Panel on Climate Change has reached the conclusion that, while the effect of urban heat islands is real, and despite the fact that other land use changes may affect climate—no surprise in either case—the effects are most important at the local and regional scale. The effect on global temperature is negligible.[42]

Evidence abounds that the warmer temperatures are related to a warmer climate: water vapor content of the atmosphere has increased, consistent with the fact that warm air can hold more water;[43] glacial ice mass and snow cover is decreasing;[44] the Arctic ice pack is thinning and shrinking;[45] permafrost is melting around the Arctic;[46] the upper 3,000 meters (9,800 feet) of the oceans are warming (absorption of carbon dioxide from the atmosphere is also acidifying the oceans, posing a threat to some marine ecosystems).[47] None of the above trends can be explained on the basis of urban heat islands or land-use changes.

Crichton, in his "Author's message" at the end of the book, makes

---

[39] David E. Parker, "Large Scale Warming Is Not Urban," *Nature* 432 (2004): 290.

[40] —, "A Demonstration That Large-scale Warming Is Not Urban," *Journal of Climate* 19 (2006): 2882–2895.

[41] Roger A. Pielke Sr. and Toshihisa Matsui, "Should Light Wind and Windy Nights Have the Same Temperature Trends at Individual Levels Even If the Boundary Layer Averaged Heat Content Change Is the Same?" *Journal of Geophysical Research* 32 (2005): L21813, doi:10.1029/2005GL024407.

[42] Kevin E. Trenberth, Philip D. Jones, Peter Ambenje, Roxana Bojariu, David Easterling, Albert Klein Tank, David Parker, Fatemeh Rahimzadeh, James A. Renwick, Matilde Rusticucci, Brian Soden, and Panmao Zhai, "Observations: Surface and Atmospheric Climate Change," in *Climate Change 2007*, 235–336.

[43] Ibid.

[44] Peter Lemke, Jiawen Ren, Richard B. Alley, Ian Allison, Jorge Carrasco, Gregory Flato, Yoshiyuki Fujii, Georg Kaser, Philip Mote, Robert H. Thomas, and Tingjun Zhang, "Observations: Changes in Snow, Ice and Frozen Ground," in *Climate Change 2007*, 337–383.

[45] Ibid.

[46] Ibid.

[47] Nathaniel L. Bindoff, Jürgen Willebrand, Vincenzo Artale, Anny Cazenave, Jonathan M. Gregory, Sergey Gulev, Kimio Hanawa, Corrine Le Quéré, Sydney Levitus, Yukihiro Nojiri, C. K. Shum, Lynne D. Talley, Alakkat S. Unnikrishnan, "Observations: Oceanic Climate Change and Sea Level," in *Climate Change 2007*, 385–432.

it obvious. *State of Fear* is no mere work of fiction. The novel is an expression of his deeply held beliefs. He says we need better science, but he expresses little short of contempt for the scientific community. Apparently the decades of scientific study of the environment have left us with no better understanding of how it works; he declares most attempts to manage natural areas a failure. Of the current consensus on global warming, he seems to view the scientists who agree with it as little more than scientific whores, manipulating data to give funding agencies the answers they want. While that does happen in the sciences—is there *any* professional field where some experts do not do such?—most researchers who seek grant money get it to ask the questions the funding agencies want answered. They do not get the money to provide the agency with cover to embrace predetermined solutions. (If the answers are already known, it is a lot less expensive to scrap the research, anyway.) A lot of scientists are driven by a childlike curiosity. Knowing the answers they are to provide before embarking on a research project would take the fun out of their work. It would destroy the joy of discovery that compels most to go into research in the first place.

Crichton is skeptical of the environmental movement, plainly saying that it is just as responsible as governments and economic interests in the exploitation of the environment. He seems to view the concern over global warming as the latest in a series of environmental alarms—going at least as far back as Thomas Robert Malthus in his 1798 *Essay on the Principle of Population*—of impending doom that never quite arrives. Malthus claimed humanity faced a population crisis in which our increasing numbers would outstrip food and other resources available, leading to widespread shortages and societal chaos. Malthus wrote:

> The power of population is so superior to the power in the earth to produce subsistence for man, that premature death must in some shape or other visit the human race. The vices of mankind are active and able ministers of depopulation. They are the precursors in the great army of destruction; and often finish the dreadful work themselves. But should they fail in this war of extermination, sickly seasons, epidemics, pestilence, and plague, advance in terrific array, and sweep off their thousands and ten thousands. Should success be still incomplete, gigantic

> inevitable famine stalks in the rear, and with one mighty blow levels the population with the food of the world.[48]

Malthus's predictions were based on a static view of population growth and resource availability; they thus failed to foresee how changes in technology would affect resource supply. So far, society has found inventive ways to boost food production. It has seemed to escape collapse into resource-deprived chaos, thus bolstering the beliefs of those who feel such concerns are overblown. Crichton, in his "message," is pretty clear about his opinion of Malthus and those like him who worry about the environment's capacity to support humanity's insatiable demand for *more*:

> I think for anyone to believe in impending resource scarcity, after two hundred years of such false alarms, is kind of weird. I don't know whether such a belief today is best ascribed to ignorance of history, sclerotic dogmatism, unhealthy love of Malthus, or simple pigheadedness, but it is evidently a hardy perennial in human calculation (*State of Fear* 570).

It is true that humanity has had a pretty good run since the dawn of the Industrial Revolution. More wealth, better technology, new ideas in political and economic philosophy: All seem to have fueled a golden age of freedom and prosperity. But, for anyone who is not ignorant of history, there are examples of the kinds of future Malthus envisioned—examples Crichton does not acknowledge. How can one explain much of the chaos of the twentieth century, such as the Russian Revolution, the rise of Fascism, or World War II, without taking into account the role of resource scarcity and resulting economic and social chaos? Given the death toll from our misadventures in the past 100 years, it seems that Malthus may have had his principles right, even if his timing was off. History is littered with the remains of civilizations that lived and ultimately died beyond their means. Some left their names. Others did not. But their ruins are a monument to the suffering of billions of our fellows who preceded, and predeceased us.

---

[48] Thomas Robert Malthus, *An Essay on the Principle of Population, as It Affects the Future Improvement of Society* (London: J. Johnson, 1798.)

Will we be the exceptions? Or will we find that the rules apply to us, too? I cannot say I am afraid. But there are times where I do get *very* nervous....

DAVID M. LAWRENCE has never decided what he will do when (if) he grows up. He is a scientist who teaches geography, meteorology, oceanography, and (sometimes) biology at the college level. He is a journalist who covers everything from high school sports to international research in science and medicine. He is a scuba diver looking for a way to make a living on the water. When not consumed with those activities, he looks at his guitars and wonders if he's too old to become a rock god. (It would help if he could actually play.) He lives in Mechanicsville, VA, with his wife, two children, and a menagerie of creatures with legs, scales, and fins.

SCIENCE COMES SECOND IN *NEXT*

Phill Jones

Like Terminal Man *years prior,* Next *was another of Michael Crichton's attempts to call out to the general public, "Hey, do you KNOW what's going on over here?" In this case, the issues were in regard to the biotechnology industry. Can your genes be patented? Can your tissues be used without your consent? What types of transgenic chimeras will humans produce? These are some of the concerns raised in* Next, *but how legitimate are they? Phill Jones looks into what's* Next.

HE NOVEL IS FICTION," Crichton says in *Next*'s disclaimer, "except for the parts that aren't."

In its e-book release of *Next*, HarperCollins Publishers included an interview with Michael Crichton. The interviewer tries to get the author to clarify just how much of *Next* is true and how much is fiction.

"It's odd but nearly everything in the book has already happened, or is about to happen," Crichton replies. "The book does look to the future a bit, particularly with regard to some transgenic animals that become important characters. But for the most part, *Next* is not really speculative fiction at all."

Oh, really?

Next is a stew of speculation; it's only lightly seasoned with true science. *Next*'s chatty apes, for instance, have a closer kinship to the surgically transformed Beast Men in H. G. Wells's *The Island of Doctor Moreau* than to bona fide genetically engineered animals.

Let's take a look at some of those parts of *Next* that aren't fiction.

A Hitchhiker's Guide to the Real Transgenic Animals

Way back in 1981, Jon Gordon and Frank Ruddle minted the term "transgenic" to describe a mouse that they had modified by inserting a foreign gene into the animal's DNA. By the early 1990s, scientists routinely produced transgenic animals with one or more foreign genes, called transgenes. Some transgenic animals did not acquire new genes; they had one or more genes deleted by genetic engineering. Contemporary transgenic animals possess any defined genetic modification.

Although transgenic animals exist, they tend to be less flamboyant than the stars of the *Next* world. Real transgenic animals don't swear in multiple human languages, or carry on civil conversations with people. Nor do they tutor kids in math. They do model, however.

The majority of transgenic animals are mice designed for scientific and medical research. Scientists may generate about 300,000 lines of transgenic mice in the near future. Many of these transgenic mice will provide experimental models to study the causes, symptoms, and potential treatments of human diseases.

In the early 1980s, Harvard Medical School researchers produced the OncoMouse®, one of the first transgenic animals and one of the few mice—other than that Mickey fellow—with a legally protected name. The OncoMouse® has a human gene that causes the animal to be highly susceptible to cancer. Scientists use the OncoMouse® to study the growth of cancers and to test treatments for breast cancer, prostate cancer, and other forms of cancer.

The OncoMouse® is not alone. Cancer researchers use many types of transgenic mice that carry one or more human genes. Other transgenic mice supply models of degenerative disorders, prion diseases, and diseases caused by bacteria and viruses. Scientists use some transgenic animals to study how organs and tissues develop and age.

While mice provide the most popular models, researchers also study human disorders in transgenic goats, pigs, and rabbits. Certain transgenic pigs, for example, have an altered growth hormone releasing hormone gene, a modification that has provided insights into Turner's syndrome, Crohn's disease, renal insufficiency, and intrauterine growth retardation.

Transgenic animals serve yet another vital purpose in drug development. Researchers use the animals to test new therapies before starting trials with human volunteers.

Some transgenic animals don't help scientists to test new drugs; they make the drugs. These "bioreactors" inhabit a field of endeavor called "pharming."

Advances in molecular biology ushered in an era where genetically engineered bacteria and yeast synthesize human proteins for therapeutic uses. Even genetically enhanced bacteria and yeast cannot make all human proteins. Some proteins require a mammalian touch to confer a certain shape for proper function or to attach sugar molecules in the correct places. These proteins can be produced in genetically engineered mammalian cells maintained in huge vats. Low yields can drive up the cost of proteins synthesized in mammalian cell cultures. For this reason, a company may turn to transgenic pigs, goats, sheep, or cattle to produce a biopharmaceutical.

During the mid-1980s, scientists interested in producing biopharmaceuticals switched their attention from metal vats to livestock mammary glands. Mammary glands seemed to offer a ready-made mechanism for producing therapeutic proteins. After all, the glands normally synthesize large amounts of complex proteins ready for secretion. Co-opting the glands might have seemed simple. But it wasn't.

Genzyme is one of the pharming pioneers. During the early 1990s Genzyme spun off a separate entity first called Genzyme Transgenics, and then GTC Biotherapeutics. GTC began work on transgenic goats that would produce in their milk human antithrombin, a protein that inhibits clotting. In August 2006, the European Commission approved GTC's biopharmaceutical for the treatment of patients with a hereditary deficiency of antithrombin. Thanks to the Commission's decision, human antithrombin became the first biopharmaceutical synthesized by a transgenic animal for commercial production.

Other organizations are working on ways to make a variety of therapeutic proteins and vaccines in the milk of transgenic goats, sheep, pigs, cows, and even rabbits. One company, Origen Therapeutics, has produced transgenic chickens that make human antibodies, which can be harvested from eggs.

For years, researchers have investigated two further applications of transgenic technology. In 1982, Richard Palmiter and his colleagues introduced transgenic mice with a rat growth factor gene. The animals loomed over their normal littermates. This might have inspired some to think about making really, really big mice. Others saw the potential to improve livestock with transgenic technology.

Modifying milk composition nourishes a number of investigations. Researchers have boosted the amounts of casein proteins in milk by producing transgenic cows with extra copies of casein genes. Casein-enhanced milk can increase the efficiency of cheese production. Other researchers have focused on reducing the lactose content of milk, which would open the market for lactose-intolerant consumers. Here, one approach is to produce a transgenic cow that secretes a lactose degrading enzyme in its milk.

Other economically important traits under investigation include improved carcass composition and enhanced meat quality. One transgenic pig carries a human insulin-like growth factor, and has larger loin mass, more carcass lean tissue, and less total carcass fat. Another type of transgenic pig produces a more healthful type of meat due to a spinach gene that synthesizes increased amounts of non-saturated fatty acids.

Genetic engineering may yield transgenic livestock with an increased resistance to diseases. One research group might have found a way to inhibit bacterial infections that cause bovine mastitis, an inflammatory reaction of the mammary gland. This disease not only harms cows, but also damages industry; mastitis creates annual losses running over a billion dollars in the United States. A research group produced transgenic cows that secrete the bactericidal enzyme lysostaphin into their milk. The genetic alteration enabled the cows to resist one of the bacteria that cause mastitis. Other disease targets include bovine spongiform encephalopathy (mad cow disease) and brucellosis.

A fourth area of investigation in the transgenic field focuses on organ transplants. Progress in the human-to-human transplantation of organs has created a severe shortage of organs available for transplant. For over forty years, xenotransplants—the transplant of tissue from a nonhuman—has been considered as a way to cope with the need for organs. At one time, primate-to-human organ transplantation seemed to offer hope. Later studies showed that organs from domesticated pigs might present the best alternative to human organs.

The human immune system creates a significant obstacle to xenotransplants. When the immune system recognizes tissue as foreign, it mounts an attack against it. Genetic engineering may provide a way to slip porcine organs under the immune system's radar. One strategy has been to produce transgenic pigs with human proteins on their cells that block an immune attack. Another tactic has been to engineer transgenic pigs that lack genes for certain pig-specific proteins.

Today, it's research models and bioreactors. The near future may see transgenic livestock with altered commercially valuable traits and transgenic pigs raised for xenotransplants.

And then, there are the transgenic animals that have a certain glow.

Critters That Glow and *Next*'s Genomic Advertising

The jellyfish has provided one of the classic transgenes, a gene that encodes a fluorescent protein. Many researchers use the *Aequorea victoria* green fluorescent protein gene as a marker to determine the efficiency of a technique for producing a transgenic animal. Sea anemone and synthetic fluorescent protein genes also have proved useful.

In *Next*, a news clip describes several unusual uses of fluorescent protein genes: the production of a glowing rabbit and glowing fish. These animals inhabit that part of *Next* that isn't fiction.

At first blush, fluorescent rabbits and fish seem to represent a frivolous use of genetic engineering. There's a little more to each story.

In 2000, the artist Eduardo Kac announced the existence of a transgenic rabbit that he had dubbed "Alba." Boasting green fluorescent protein (GFP) genes, Alba had a dual identity. The albino rabbit appeared white with pink eyes in daylight. A blue light transformed Alba into a brightly glowing, green rabbit. Sound unlikely? The artist released a photo to prove it.

Kac reportedly claimed that he had commissioned Louis-Marie Houdebine and other scientists at France's National Institute of Agronomic Research to fashion Alba for him. He thought that his "transgenic art" rabbit would spark worldwide controversy about the ethics of using genetic engineering in this way. Alba did breed controversy.

"When E. Kac visited us, we examined three or four GFP rabbits,"

Houdebine told *Wired Magazine* in 2002. "He decided that one of them was his bunny, because it seemed a peaceful animal."

Houdebine explained that researchers had generated the GFP rabbits as models to investigate the fate of embryonic cells in developing embryos. He denied that they had produced Alba for the artist. Houdebine also asserted that Kac's famous photo of a uniformly glowing rabbit owes more to special effects than genetics. While the rabbit's eyes and ears appear green under ultraviolet light, its fur doesn't glow.

What about those glowing fish? They exist. In fact, Yorktown Technologies markets GloFish® fluorescent zebra fish for home aquariums. Take your pick: Starfire Red™, Electric Green™, or Sunburst Orange™.

The fish did not begin as a novelty item. In 1999, Zhiyuan Gong and his colleagues at the National University of Singapore made transgenic zebra fish that had green fluorescent protein. As an encore, they produced yellow, orange, and red transgenic zebra fish. Their goal was to devise genetically engineered fish that would signal the presence of certain pollutants in water. For example, the presence of small amounts of cancer-causing polychlorinated biphenyls (PCBs) might cause a black and silver zebra fish to emit green light when radiated with blue light.

The denizens of the *Next* world devise a less practical way to couple transgenic technology with fluorescent proteins, bioluminescent proteins, and pigments: genomic advertising. Gavin Koss introduces the concept with a presentation at his London advertising agency.

"The natural world," Koss begins, "is entirely without advertising. The natural world has yet to be tamed. Colonized by commerce." Koss then launches into his pitch for genetically engineering animals into living billboards.

Like many of the villains in *Next*, Koss feels unbound by rational thought. It's not enough to unleash transgenic advertisers; they must be more genetically fit than their natural counterparts. This way, Koss explains, their clients' transgenic animals will out-compete unmodified animals, driving these message-deficient creatures into extinction.

"We are entering the era of Darwinian advertising!" Koss cries out. "May the best advert win!"

Koss's audience is unimpressed. Somebody has already begun genomic advertising. Unknown to Koss, a turtle has been spotted in Costa Rica, a turtle bearing a flashing corporate logo on its shell.

Is genomic advertising just around the corner?

"I know of no genetically engineered messages of any type on animals," says Professor C. Neal Stewart, Jr., of the University of Tennessee. "I'd be astounded if such a thing were to be done in the next twenty-five years."

Stewart, an expert in the various uses of green fluorescent protein and genetic engineering, suggests that regulatory issues present a significant barrier. He doesn't think that government regulators would allow an unrestricted release of transgenic adverts into the environment, especially animals engineered to be fitter than their natural counterparts.

Genomic advertising also has technical barriers. "There is no marker gene suitable for that sort of thing yet," Stewart says. Fluorescent proteins would not work; they require excitation light of certain wavelengths. Current technology cannot produce a bioluminescence that would be sufficiently bright. Chromoproteins might be used to change the color of skin, but devising a method for delivering the genes to make a permanent change in the skin presents a tough hurdle.

"I'm not sure," Stewart says, "that the public would accept such a thing."

Even in the *Next* world, the public greet genomic advertising with outrage.

How Do They Do That?

Next's Mark Sanger, a wealthy aspiring artist, buys stacks of textbooks to learn about genetic engineering. He scans figure legends that mention LoxP, Cre recombinase, lentiviral vectors, homologous recombination, and the like. Without tackling the text itself, Sanger gives up. We can do better than that.

Remember those large transgenic mice with rat growth hormone genes? Scientists produced the animals by injecting DNA into a very young embryo. It works like this.

After a sperm cell fuses with an egg cell, the cell contains egg and sperm nuclei, called pronuclei. The pronuclei fuse to create one nucleus with a full set of chromosomes. Before they fuse, a fragment of DNA that includes the transgene can be injected into one of the pronuclei. The foreign DNA becomes integrated into the chromosomes, and the single cell continues embryonic development.

DNA microinjection remains a popular method for producing transgenic rodents. The technique has also been used to make transgenic rabbits, sheep, swine, goats, and cows. The efficiency of the technique varies greatly among species due to differences in a capacity to integrate foreign DNA.

Virus-mediated delivery offers an alternative to DNA microinjection. A virus is basically a DNA or RNA molecule wrapped in proteins. Since a virus travels light, it lacks the wherewithal to reproduce itself. Instead, the virus hijacks the protein and nucleic acid synthesis machinery of a host cell.

A retrovirus, for example, is an RNA virus that infects mammalian cells. To reproduce itself, the virus slips its RNA molecule into a cell. The cell synthesizes a DNA copy of the viral RNA molecule, and the viral DNA molecule integrates into the host cell's DNA. Now enthroned within a chromosome, the viral DNA directs the cell to synthesize more viral RNA molecules and proteins to mass-produce new viruses.

Scientists retrofit retroviruses into transgene delivery systems. They delete genes required for virus reproduction and add one or more new genes. A recombinant retrovirus can efficiently deliver transgenes to young embryos. Researchers have used retroviral vectors to generate transgenic chickens, pigs, sheep, and cows.

Techniques like DNA microinjection and virus-mediated delivery of a transgene suffer from a serious limitation: the transgene becomes integrated into host DNA in a random manner. As Crichton explains, the gene must be incorporated correctly in the DNA of the recipient animal.

"Sometimes the new gene was incorporated backward," Crichton says, "which had a negative effect, or none at all. Sometimes it was inserted into an unstable region of the genome, and triggered lethal cancer in the animal" (169).

Randomly inserting a transgene into a genome can become the molecular equivalent of tossing a monkey wrench into a piece of intricate machinery. Disruption of normal gene function by a transgene—insertional mutagenesis—can trigger disastrous results.

In 2003, for instance, scientists reported a tragedy in a gene therapy trial. Two children developed cancer after receiving cells treated with a recombinant virus. The viral DNA had become inserted into the cell's DNA in a way that promoted the growth of cancer cells.

Researchers have devised a way to control the insertion of a transgene. *Next*'s Mark Sanger reads about it while flipping through textbooks. The method uses homologous recombination, or DNA crossover.

In DNA crossover, two DNA molecules line up and exchange DNA fragments. This DNA swap occurs when the two DNA molecules contain two stretches of identical nucleotide sequences.

Here's an example. Suppose that a chromosome in a sheep cell includes the following: "nucleotide sequence 1—Gene x—nucleotide sequence 2." To induce a DNA crossover, a researcher treats the cell with a DNA molecule that includes "nucleotide sequence 1—Gene y-nucleotide sequence 2." The cell's machinery may use the identical nucleotide sequences like handles to swap pieces of DNA. Any nucleotide sequences that lie between these handles get moved as well. If a swap occurs, then the sheep chromosome will contain Gene y, rather than Gene x.

Other approaches for gene swapping are available. Mark Sanger learns about the bacteria-derived Cre-LoxP system for provoking a gene exchange. This complicated tactic currently has a low success rate.

A researcher may use DNA crossover to replace an active gene with an inactive one, a technique often called knock-out. Transgenic knock-out mice have been used to study cancer, heart disease, diabetes, arthritis, and other disorders.

If gene swapping replaces one gene for an active gene, the process is called knock-in. The new gene may be a mutant version of the replaced gene. As an example, researchers generated a mouse model for sickle cell disease by swapping a normal hemoglobin gene with a mutant hemoglobin gene.

The production of knock-out or knock-in mice can require a combination of genetic engineering, DNA crossover, and stem cell technology. It is not for the easily frustrated or impatient.

As Crichton says, "The successful injection of transgenes was a tricky business, and required dozens, even hundreds, of attempts before it worked properly" (168).

And yet, *Next* scientists manage to genetically engineer talking apes, as well as a parrot that not only converses with humans, but also tutors students in math.

Look Who's Talking: Two Apes and a Parrot Who Does More Than Ape Humans

Next features two genetically altered apes: an orangutan that profusely swears in several languages, and a chimpanzee. We learn little about the orangutan, but Dave the chimp plays a major role.

Researcher Henry Kendall, who is responsible for Dave, adopts the chimp. Despite his simian heritage, Dave fits into the Kendall family.

"When he went outside, he wore a baseball cap, which helped his appearance a lot," Crichton says. "With his hair trimmed, wearing jeans and sneakers and a Quicksilver shirt, he looked much like any other kid" (215).

Dave can carry on a conversation, write, and read. He even adjusts—for a while—to life in an elementary school. Teachers, struggling to cope with school kids, fail to spot Dave as a nonhuman primate.

It's true that humans and chimpanzees are close. Chimps are humans' closest living relatives.

Ancestors of humans and chimpanzees went their separate ways around 5 to 6 million years ago. According to one line of thought, environmental changes forced the split. At that time, the planet faced global cooling that locked up enormous amounts of water in glaciers. In equatorial East Africa, forest dwelling apes faced a dilemma: a drought that diminished forests. Some apes toughed it out in the trees and gave rise to the lineage that led to chimpanzees. Others left the forests, facing predators and new challenges in open lands. These apes gave rise to the human lineage.

Perhaps to mollify his wife about Dave, Henry Kendall tells her that the genome of the chimpanzee is nearly identical to that of a human. He's correct—in a way.

In 2005, the Chimpanzee Sequencing and Analysis Consortium reported that human and chimp genomes differ by one to two percent. This may seem like an insignificant amount, but it translates into over 35 million changes in nucleotide sequences.

Further studies revealed more differences between humans and chimps. As the human and chimp lines evolved, mutations inactivated ancient genes. Today, humans lack genes that chimpanzees possess and vice versa. Even when human and chimp cells have the same active genes, the cells may synthesize different amounts of proteins or assemble proteins differently.

Do scientists understand the genetic bases for the human linguistic facility, complex abstract thought, and other features that distinguish humans and chimpanzees? Can they use genetic engineering to produce a transgenic chimpanzee with humanlike abilities in speech and reason?

"My short answer is 'no,'" says Matthew Hahn, an expert in comparative genomics at Indiana University. "The main reason for this is that there are not just one or two mutations that distinguish humans and chimpanzees, or even their language abilities, but dozens to hundreds."

Hahn cites one example. "The gene MYH16 is completely missing in humans," he says, "and has allowed our jaw and cranium to take on a much different shape than our closest relatives."

The MYH16 gene, Hahn says, "is only one of the morphological—not to mention neurological, behavioral, and sensory—differences that makes it possible for us to use language. In addition, because we have yet to identify most of the causal mutations underlying these differences, it would be impossible to know which pieces of DNA would be needed to give a chimpanzee language abilities."

Bruce Lahn, a professor of human genetics at the University of Chicago, also sees technological hurdles in the genetic engineering of apes with human reasoning abilities.

"We don't yet understand what genomic differences between human and ape are responsible for cognitive differences between the species," Lahn says. "Even if we did, the amount of genetic engineering may be far too daunting than what the current technology can offer."

So, just how did Henry Kendall produce Dave? Apparently, Kendall gives a here-goes-nothing shrug and chucks all of his genes into a chimpanzee embryo.

Although Kendall does not seem to realize it, Dave—an amalgamation of undefined genetic alterations—is not a "transgenic" animal. Instead, Dave's jumble of chimp and human genomes makes him a hybrid, like the gorilla-human hybrid in Maureen Duffy's *Gor Saga* (1981).

Could somebody make a hybrid by blending human and chimp genomes?

Anthony W. S. Chan of Emory University's School of Medicine says that it's possible to produce an animal by replacing the nucleus of an oocyte from one species with the nucleus of a cell from a different species. The technique generates an animal that's essentially a clone of the animal that provided the donor nucleus.

So, a nucleus *swap* can produce an animal. Yet Kendall begets Dave by *adding* the genetic contents of a second nucleus.

"As far as I am aware," Chan says, "no animal has been created with both a recipient oocyte nucleus and a donor nucleus from two different species."

Chan knows about transgenic nonhuman primates. He was a member of the group—and so far, the only group—to report the production of a transgenic rhesus monkey.

To generate the transgenic animal, the researchers treated rhesus monkey egg cells with a modified retrovirus that included a gene for a green fluorescent protein. The scientists were not trying to produce a glowing green monkey. The green fluorescent protein only served as a readily detectable marker to evaluate the success of the procedure. The long-term goal of genetically modifying rhesus monkeys is to generate primate models of human diseases that cannot be studied in other species.

What about transgenic apes? Where are the transgenic chimps and orangutans?

"As far as I know," Chan says, "I don't think that anyone has made a transgenic chimpanzee or orangutan." This is not because it is impossible, he says, but rather, no one has had a sufficient reason to try.

If transgenic apes are not just around the corner, how about a savvy transgenic parrot akin to *Next*'s Gerard? After all, Irene Pepperberg's studies with the African Grey parrot Alex indicate that the bird not only has learned a vocabulary of at least 100 words, but also can respond to a question using reason, not mimicry. Can genetic engineering boost a parrot's natural abilities and transform it into a Gerard-like, self-aware conversationalist?

"I do not think it is possible to produce a transgenic parrot with all the humanlike abilities in speech and reason," says Erich D. Jarvis. His lab at Duke University Medical Center investigates the neurobiology of vocal communication by combining behavioral, anatomical, and molecular biological techniques.

"I do think it will be possible," Jarvis says, "to produce transgenic parrots with more abilities than they already have, including producing longer sequences of speech with meaning." He cautions, however, that this will require genetically engineering connectivity within the parrot brain.

A Patently Evil Practice?

Much of *Next* concerns allegedly improper responses of the legal system to advances in biotechnology, such as gene patenting. During an interview, Crichton condemned gene patenting as an "obstructive and even dangerous" practice. This outlook comes across in *Next*, where it's the villains, such as Robert Bellarmino, who favor the patenting of genes.

The *Next* world proponents of gene patenting do not overlook an opportunity to mischaracterize the nature of a gene patent.

"The notion that someone owns part of the human genome strikes some people as unusual," says Bellarmino.

It's not only unusual; it's incorrect. Nobody owns parts of the human genome.

The federal government grants three types of patents. The utility patent is the most significant type of patent in the field of biotechnology. To obtain a utility patent, an inventor files a patent application with the U.S. Patent and Trademark Office. Here, a patent examiner scrutinizes the application to decide whether the document meets legal requirements and justifies the inventor's patent claims.

For example, an examiner determines if the claimed invention qualifies as a process, machine, manufacture, or composition of matter. A person cannot patent a law of nature, a physical phenomenon, or an abstract idea. Can you patent your discovery of a preexisting object? No. But you may be able to patent an application of your discovery.

In exchange for a patent, a patent applicant must fully describe the claimed invention. A patent application must contain a sufficiently detailed description of the invention to enable others to make and to use the invention. A patent application must reveal what the inventor considers to be the best method for carrying out the invention and must describe a specific use for a claimed invention.

The patent examiner also determines if the claims include something known to the public, or something that could have been readily deduced from publicly available information by a person knowledgeable about the relevant technological field. In either case, the examiner will reject the claims.

After overcoming these hurdles, the patent office grants a patent. Now, the patent owner has the right to exclude others from making, using, offering for sale, or selling the patented invention within the U.S., or importing the patented invention into the U.S. Today, these patent rights last for a period of twenty years from the filing date of the patent application. Since the patent examination process requires two, three, four, or more years, a patent term tends to be much less than twenty years.

Next correctly shows that the idea of patenting genes and gene fragments has sparked controversy at least since the early 1990s when the National Institutes of Health filed three patent applications that covered over 6,000 DNA fragments. James Watson, who directed the NIH genome project at that time, denounced the plan to patent DNA bits as "sheer lunacy," contending that "virtually any monkey" could crank out the nucleotide sequences using automated sequencing machines. The NIH faced an onslaught of criticism from academia and the biotech industry, as well as the possibility of losing international collaboration in mapping the human genome. The organization withdrew the patent applications.

Since then, federal courts have made it clear that claims on pieces of DNA with unknown use cannot meet patentability requirements. What about a claim on a complete gene with a credible use?

A naturally occurring molecule cannot be patented even if the patent applicant were the first to discover the existence of the molecule.

On the other hand, a purified, isolated, or altered form of a naturally occurring molecule may be patentable.

An early court case on gene patenting—early as in 1989—concerned claims to human erythropoietin, a protein that became a blockbuster therapeutic for Amgen, Inc. A federal judge explained that the nucleotide sequence encoding human erythropoietin is a non-patentable natural phenomenon. However, Amgen claimed an isolated and purified nucleotide sequence that encodes human erythropoietin, which can be patented.

The number of U.S. patents on DNA molecules rose sharply from the mid-1990s through 2001, bringing a tempest of dissent. Critics urged that the patent land grab could throttle research and development.

Yet the sky didn't fall. Recent studies of universities and industry show that feared ill effects of gene patenting have been relatively uncommon. In one survey, for example, only 1 percent of U.S. academic biomedical researchers reported that they had to delay a project due to patent entanglements.

As *Next* points out, gene patenting has hampered efforts to develop certain genetic tests. The most notorious case concerned the 1997 enforcement of a patent on isolated DNA molecules encoding BRCA-1, a protein linked to susceptibility for breast and ovarian cancer. It remains the most controversial case.

Alex's Adventures Through the Looking-Glass

A significant portion of *Next* follows the misfortunes of Alex Burnet, her father, and her son. Crichton takes readers on a journey into a reality-deficit world terrorized not by Jabberwocks, but by evil lawyers and greedy biotech executives.

The Burnet saga begins when Frank Burnet discovers a growth in his abdomen. He seeks help from Dr. Michael Gross at the University of California, Los Angeles Medical Center. Gross takes tissue samples, runs a battery of tests, and pronounces a diagnosis of acute T-cell lymphoblastic leukemia.

Surgery and chemotherapy appear to cure Burnet. Nevertheless, Gross insists that Burnet return for follow-up exams and removal of tissue for testing. This continues for four years. Burnet eventually learns that Gross and the Regents of the University of California sold the rights to Moore's cells to BioGen. Moore's cells are worth billions of dollars, because they churn out massive amounts of valuable proteins called cytokines.

Burnet sues Gross and UCLA, but loses. The judge concludes that Moore has no rights in tissue removed from his body.

After someone sabotages BioGen's cell lines, the company's villains decide to replace the cells by kidnapping Burnet or a family member and subjecting that person to invasive procedures to extract tissue. The BioGen folks arrive at a fantastic interpretation of the judge's ruling: Burnet no longer owns his own cells—even those within his body. Since Burnet's daughter Alex and her son Jamie have the same cells, BioGen can remove the cells from them. They sic bounty hunters after Alex and her son, an action they deem legal, because their targets are walking around with stolen property.

Yikes!

From a legal standpoint, the scheme is ludicrous. No sane interpretation of law would back the actions of these jokers. They'll get cells alright—either padded or with iron bars.

The science also fails to stand up to scrutiny.

Burnet seems to have his valuable cells as a reaction to acute T-cell lymphoblastic leukemia. Environmental exposure, possibly to a virus or chemical, seems to play an important role in initiating the disease. Even if Burnet had an identical twin—with an identical genetic makeup—nobody could guarantee that Burnet's twin would have the valuable cells.

But let's assume that Burnet's cells are entirely due to his genetic makeup. Why would his daughter Alex have the same cells? She did get half of her genes from her mother. For that matter, why would Alex's son Jamie produce the cells?

The Burnet chronicles unfold along a bizarre path. Yet the saga is based upon a true court case: *John Moore v. The Regents of the University of California*, decided by the Supreme Court of California in July 1990.

"*Moore* was a case very much like yours," Alex tells her father. "Tissues were taken under false pretenses and sold. UCLA won that one easily, though they shouldn't have."

Well, it wasn't quite like that.

In their decision, the California Supreme Court presented the allegations of Moore's complaint as follows. In October 1976, John Moore visited the UCLA Medical Center after he received a diagnosis of hairy-cell leukemia. Dr. David Golde hospitalized Moore, ran tests, confirmed the diagnosis, and recommended a course of treatment, including a splenectomy to slow the disease's progress. Without requesting Moore's permission, Golde arranged a transfer of part of Moore's spleen to a research unit.

At Golde's direction, Moore returned several times to the UCLA Medical Center between November 1976 and September 1983. On each visit, Golde withdrew samples of blood and other tissues. Moore did not know that Golde and others transformed Moore's white blood cells into a cell line that overproduced lymphokines.

Golde and a researcher assigned their rights as inventors to the Regents of the University of California. The Regents filed for and obtained a U.S. patent on the cell line. With the Regents' help, Golde negotiated agreements with two biotech companies for commercial development of the cell line and products synthesized by the cells.

Moore discovered that his cells had been spun into a golden cell line. He sued the Regents, Golde, and researchers on thirteen grounds.

Two trial court judges declared that Moore had not presented a case that could move forward. Moore appealed.

The Court of Appeal reversed the trial judges. The appellate court decided that Moore stated a cause of action for conversion, a tort that protects against interference with ownership interests in personal property. According to Moore's theory, the defendants' unauthorized use of his cells constituted a conversion, because Moore continued to own his cells even after the cells had been removed from his body. Moore also claimed a proprietary interest in all products synthesized by his cells and by the patented cell line.

The next appeal landed the case in the California Supreme Court. The highest court of the state ventured into alien territory. Legal precedent did not exist for a claim of conversion liability arising from the use of human cells in medical research. Like Indiana Jones, they would make it up as they went along.

First, the court reviewed existing law and could find no support for

Moore's position. Even if Moore had an ownership interest in excised cells, he could not collect on the valuable cell line, the court decided, because the cell line was "factually and legally distinct" from the cells taken from Moore's body.

Although the court could have extended conversion liability, it declined to do so. In the court's view, this extension of conversion theory would harm innocent parties. Since conversion is a strict liability tort, it would impose liability on anyone who used the cells, including those who knew nothing about any impropriety or illegality in procuring the original cells.

The court offered another reason why it declined to make new law: Moore had an avenue to address any wrongs. He could sue to enforce a physician's obligation to disclose any research and economic interests that may affect judgment.

"The fiduciary-duty and informed-consent theories protect the patient directly," the court majority wrote, "without punishing innocent parties or creating disincentives to the conduct of socially beneficial research."

Not everyone was on board with the decision. Two judges found support for Moore's claim of conversion.

Almost two decades later, the debates continue about how, or if, an individual who supplies cells for research should be compensated for the contribution. So far, bounty hunters, kidnapping, and forcible extraction of tissues have not heightened the ethical dilemma.

A Light at the End of the Yellow Brick Road

Grotesque creatures inhabit *Next*. Many of the nonhumans are odd, as well.

Clear away the book's more fantastic elements. Chase the talking apes back to the trees. Short-circuit that nattering parrot's brain with an insolvable math problem. Pull the plug on the glowing advertising animals. Then, shine a light on *Next*, one bright enough to send the giant pet cockroaches and serpentine lawyers scurrying for cover.

And there, on the shelf behind the gene therapy that cures addiction by inflicting rapid, fatal aging, you can find a message about genetic engineering. It's about continuing scientific research—with caution.

Today, the major differences between genetically modified species and their wild counterparts have nothing to do with the application of modern science. Humans have introduced these alterations during thousands of years of domestication.

Advances in genetic engineering allow humans to accelerate genetic modification, to engineer changes in a much shorter time frame. This capability raises the question: Just because we can, should we?

PHILL JONES is a freelance writer. His articles have appeared in *History Magazine*, *Nature Biotechnology*, *Forensic Magazine*, *The World Almanac and Book of Facts*, and other publications. *Futurismic ezine*, *Far Sector SFFH*, *Apex Science Fiction & Horror Digest*, and *Women's World* have published his short stories. He also teaches an online forensic science survey course for writers.

A former associate professor of biochemistry, he once earned a living as a patent attorney, specializing in biotech inventions. He never had the opportunity to work with a talking transgenic animal that could explain why it met patentability requirements.